Jesús Adrián López
Angelica Judith Granados López
Luis Steven Servín González

microRNAs: a powerful tool for cervical cancer therapy and diagnosis

Jesús Adrián López
Angelica Judith Granados López
Luis Steven Servín González

microRNAs: a powerful tool for cervical cancer therapy and diagnosis

microRNAs: the future for cervical cancer therapy and diagnosis

LAP LAMBERT Academic Publishing

Impressum / Imprint

Bibliografische Information der Deutschen Nationalbibliothek: Die Deutsche Nationalbibliothek verzeichnet diese Publikation in der Deutschen Nationalbibliografie; detaillierte bibliografische Daten sind im Internet über http://dnb.d-nb.de abrufbar.
Alle in diesem Buch genannten Marken und Produktnamen unterliegen warenzeichen-, marken- oder patentrechtlichem Schutz bzw. sind Warenzeichen oder eingetragene Warenzeichen der jeweiligen Inhaber. Die Wiedergabe von Marken, Produktnamen, Gebrauchsnamen, Handelsnamen, Warenbezeichnungen u.s.w. in diesem Werk berechtigt auch ohne besondere Kennzeichnung nicht zu der Annahme, dass solche Namen im Sinne der Warenzeichen- und Markenschutzgesetzgebung als frei zu betrachten wären und daher von jedermann benutzt werden dürften.

Bibliographic information published by the Deutsche Nationalbibliothek: The Deutsche Nationalbibliothek lists this publication in the Deutsche Nationalbibliografie; detailed bibliographic data are available in the Internet at http://dnb.d-nb.de.
Any brand names and product names mentioned in this book are subject to trademark, brand or patent protection and are trademarks or registered trademarks of their respective holders. The use of brand names, product names, common names, trade names, product descriptions etc. even without a particular marking in this work is in no way to be construed to mean that such names may be regarded as unrestricted in respect of trademark and brand protection legislation and could thus be used by anyone.

Coverbild / Cover image: www.ingimage.com

Verlag / Publisher:
LAP LAMBERT Academic Publishing
ist ein Imprint der / is a trademark of
OmniScriptum GmbH & Co. KG
Heinrich-Böcking-Str. 6-8, 66121 Saarbrücken, Deutschland / Germany
Email: info@lap-publishing.com

Herstellung: siehe letzte Seite /
Printed at: see last page
ISBN: 978-3-659-76733-3

Zugl. / Approved by: Zacatecas, Autonomic University of Zacatecas, Diss., 2015

Index

1. Introduction 1
2. MiRNAs families altered in cervical cancer 4
3. Cell-signaling pathways regulated by members of miRNA families expressed in clusters. 4
3.1 Regulation of PI3K-AKT and MAPK axis by miR-133b~206 cluster 4
3.2 Regulation of CUL5, NOTCH, TNKS2, PTEN-PI3K-AKT axis by miR-17~92 and miR-106A~363 clusters..6
3.3 Regulation of uPA-PLG-MMP and JAG-NOTCH axis by miR-23b~27b~24-1 cluster 9
3.4 Regulation of HSP47, YY-1 and CDK6 axis by miR-29a~29b-1 and miR-29b-2~29c clusters...11
3.5 Regulation of P18Ink4c and JAG-NOTCH-uPA axis by miR-34a................12
3.6 FOXO1-cullin-rings axis are regulated by the cluster miR183~96~182 12
3.7 Regulation of PI3K-AKT and MAPK axis by miR-let-7c~99a, miR-125a~let-7e~99b and miR-100~let-7a-2 clusters 13
3.8 Regulation of PI3K-AKT and MAPK axis by miR181a-1~181b-1 and miR181a-2~181b-2 clusters 14
3.9 Regulation of MEK, JNK and Bcl2-l2 axis by miR-214~199a-2 cluster.......15
3.10 Regulation of Cyclin D and AKT axis by miR-302s cluster........................17
4. Altered miRNAs Expression in Cervical Carcinomas 17
5. miRNAs Implicated in Cervical Cancer Progression......................................25
6. miRNAs Regulated by HPV Oncoproteins 26
7. Construction of a Multistep Model of Carcinogenesis by Expression of miRNAs and Their Targets 28
7.1. MicroRNAs Misregulated in Step 1 28
7.2. MicroRNAs Misregulated in Step 2 35
7.3. MicroRNAs Misregulated in Step 3 37
7.4. MicroRNAs Misregulated in Step 4 37
8. Conclusions 40
Author Contributions...44
Conflicts of Interest 44
Acknowledgments 44
References 44

microRNAs: a powerful tool for cervical cancer therapy and diagnosis

Jesús Adrián López[1,2,*], Angelica Judith Granados López[1], Luis Steven Servín González[1], Lucrecia Carrera Quintanar[2], Flor de María Trejo Medinilla[3], Hiram Hernández López[4] and Rosalinda Gutiérrez Hernández[5].

[1] Laboratorio de microRNAs, Unidad Académica de Ciencias Biológicas, Universidad Autónoma de Zacatecas; Av.Preparatoria S/N, Zacatecas 98066, México; E-Mail: agranadosjudith@gmail.com

[2] Área Académica de Ciencias Básicas, Doctorado de Ciencias Básicas, Universidad Autónoma de Zacatecas; Av.Preparatoria S/N, Campus II, Zacatecas 98066, México

[3] Laboratorio de Biología Molecular, Unidad Académica de Ciencias Químicas, Universidad Autónoma de Zacatecas; Carr. Zacatecas-Guadalajara Km. 6, ejido La Escondida, Zacatecas 98160; e-mail: tmedinillasm@yahoo.com.mx.

[4] Laboratorio de Síntesis e Ingenieria de Reacciones Químicas, Unidad Académica de Ciencias Químicas, Universidad Autónoma de Zacatecas; Carr. Zacatecas-Guadalajara Km. 6, ejido La Escondida, Zacatecas 98160; e-mail: tmedinillasm@yahoo.com.mx.

[5] Laboratorio de Etnofarmacología, Unidad Académica de Ciencias Químicas, Universidad Autónoma de Zacatecas; Carr. Zacatecas-Guadalajara Km. 6, ejido La Escondida, Zacatecas 98160; e-mail: rosalindagh@hotmail.com

* Author to whom correspondence should be addressed; e-Mail: jalopez@uaz.edu.mx; Tel.: +52-492-149-2648.

Abstract: Tumor cells have developed advantages to acquire hallmarks of cancer like apoptosis resistance, increased proliferation, migration, and invasion through cell signaling pathways misregulation. The sequential activation of genes in a pathway is regulated by miRNAs. Close to 70 microRNAs (miRNAs) have been implicated in cervical cancer up to now, nevertheless it is unknown if aberrant miRNA expression causes the onset of cervical cancer. Lost or gain of miRNA expression could activate o repress a particular cell axis. It is well known that aberrant miRNA expression is well recognized as an important step in the development of cancer. miRNAs are grouped in clusters and may have similar functions, such is the case of clusters with anti-oncomiRs 23b~27b~24-1, miR-29a~29b-1, miR-29b-2~29c, miR-99a~125b-2, miR-99b~125a, miR-100~125b-1, miR-199a-2~214 and miR-302s and oncomiRs function of miR-1-1~133a-2, miR-1-2~133a-1, miR- 133b~206, miR-17~92, miR-106a~363, miR183~96~182, miR-181a-1~181b-1 and miR- 181a-2~181b-2 that regulated MAPK, PI3K-AKT, NOTCH, proteasome-culling rings, and apoptosis cell signaling. In the progression of cervical cancer there are three well- established steps to reach cancer that we used in the model proposed here. The first step of the model comprises the gene changes that occur in normal cells to be transformed into immortal cells (CIN 1), the second comprises immortal cell changes to tumorigenic cells (CIN 2), the third step includes cell changes to increase tumorigenic capacity (CIN 3), and the final step covers tumorigenic changes to carcinogenic cells. Altered miRNAs and their target genes are located in each one of the four steps of the multistep model of carcinogenesis. In this work we point out the pathways regulated by families of miRNAs grouped in 20 clusters and we propose a cervical multistep model of carcinogenesis. Pathways regulated by miRNA families in combination with carcinogenesis model will increase the knowledge of cervical cancer therapeutic, diagnostic and prognostic methodologies design.

Keywords: miRNAs; clusters; families of miRNAs; cell signaling pathways; cervical intraepithelial neoplasia (CIN); cervical cancer

1. Introduction

Cervical cancer is one of the most frequent diseases in the world, and the second type of cancer that kills most women worldwide, with an estimated global incidence of 470,000 new cases and over 200,000 deaths per year [1]. One of the first events in cervical cancer development is the infection with human papilloma virus (HPV). HPV is associated with benign and malignant cervical lesions infecting mucosa and epithelial surfaces of the cervix. HPV replication occurs exclusively in squamous stratified epithelium such as the epidermis and mucous membranes[2,3] and is dependent on the cellular differentiation state and abundance of important proteins like transcriptional factors, polymerases, splicing factors, and an RNA processing machinery[3].

HPV infection results in the expression of viral proteins that change transitory normal cell functions such as proliferation and differentiation. Some HPVs, such as HPV16 and HPV18, are associated with oncogenesis and are therefore considered "high risk" (HR) viruses. HR viral E7 oncoprotein interacts with retinoblastoma (Rb) protein family members and permits G1 to S transition through the transcriptional factor E2F release from Rb, its regulatory protein [4]. Target promoter E2F binding activates transcription of several genes involved in proliferation [2]. In addition, HR-HPV E6 protein interacts with p53 and induces its degradation via ubiquinitation, resulting in p53-null phenotype abrogating apoptosis and cell cycle checkpoints [3,5]. Besides misregulation of coding genes, HR-HPV oncoprotein expression causes dysregulation of non-coding genes like microRNAs (miRNAs).

Cancer is a genetic complex pathology that involves coding gene and non-coding gene abnormalities [6]. Normal cellular proliferation is regulated by proto-oncogenes promoting proliferation and is balanced by their counterpart, tumor gene suppressors, that inhibit cellular proliferation. Mutations that increase proto-oncogen activity convert these genes into oncogenes leading to tumor cell proliferation. Normally these genes are growth factors, growth factor receptors, transductional signaling proteins, and DNA binding proteins [7]. On the other hand mutations that inactivate tumor gene suppressors liberate the genetic brake, and thereby potentiate tumor cell proliferation. However, for tumor progression, both of these genes have to be affected. Cellular proliferation is not an autonomic event; it obeys intercellular communication, ensuring normal tissue integrity. Examples of intercellular signals are contact inhibition and anchorage dependent growth, which are both hallmarks of normal cells [8]. Recently, an interchange of miRNAs by cells was reported, suggesting another way of communication [9,10]. With the discovery of miRNAs, the concepts of oncogenes and tumor gene suppressors have expanded. miRNAs regulating oncogenes are known as anti-oncomiRs, and miRNAs inhibiting tumor gene suppressors are known as oncomiRs [6,11]. One of the most important features about miRNAs is that they regulate several mRNA targets [12,13], permitting the analysis of several coding genes with just one miRNA [14].

MicroRNAs are noncoding regulatory RNAs 19–25 nucleotides (nt) in size that are produced by RNA polymerase II (pol II) and III (pol III) derived from transcripts of coding or noncoding genes. Many miRNAs are tissue-specific or differentiation-specific, and their temporal and lived expressions modulate gene expression at the post-transcriptional level by base pairing with complementary nucleotide sequences of target mRNAs[15,16]. Depending on the degree of sequence complementarity binding of miRNAs to target mRNA, they inhibit protein translation and/or degrade target mRNA[13].

Bioinformatics prediction shows that each miRNA targets more than 100 RNA transcripts and up to one-third of the total number of human mRNAs is regulated by these non-coding genes[17,18]. Therefore, the actions of miRNAs exert profound effects on gene expression in almost every biological process. Proliferation, anchorage independent growth, apoptosis, migration, and invasion are regulated by miRNAs [19]. In fact, restoration of miRNA expression or miRNA inhibition alters cellular processes [19,20]. Therefore, miRNAs are a powerful tool for gene therapy, prognosis, and diagnosis of malignant diseases.

miRNA expression affected by HPV specifically occurs as consequence of cervical cancer [21], some others are altered independently of HPV infection as a cause of cervical cancer. However, it is too soon to distinguish between the involvement of miRNAs as a consequence and/or a cause of cancer, but it is a fact that they orchestrate gene profile changes to induce carcinogenesis [22,23]. Since each miRNA reflects more than 100 coding genes regulated in cancer progression they are ideal genes to make a model of multistep carcinogenesis. Even though most of the actual papers compare normal *versus* carcinoma state and just a few have evaluated different neoplasia states in the cervical cancer process, with these data we propose a model to explain the cervical cancer progression based on miRNAs expression and their target coding genes.

Cellular paths are controlled through a variety of mechanisms, since extracellular and intracellular signals carried out by proteins, nucleotides, carbohydrates, ions like Ca2+, etc. A novel mechanism of cell path regulation is yielded by microRNAs, as small RNA molecules that actively inhibit expression of all kind of proteins. MiRNAs characteristics such as regulation of a great number of proteins by each miRNA, confers these little molecules great impact on cellular fate through cell pathway regulation. Reviewing how miRNAs families expressed in clusters regulated cell paths signaling will increase the knowledge of cervical cancer progression giving advantages for therapeutic, diagnostic and prognostic methodologies design. Several works are enriching this knowledge by the generation of experimental evidence of miRNAs participation in regulation of cell paths. Therefore coupling the information available we propose cell paths affected through miRNAs in cervical cancer. The advantages generated by gain or loss of function in miRNAs expression on cancer development [19,21] could be the result of the affectation of one or more members of the family or cluster. This work shows several miRNAs misregulated that belong to miRNAs families and clusters (Table 1) suggesting a complex system of regulation affecting genes and cellular processes. With these data we propose a model to explain the cervical cancer progression based on miRNAs expression and their target coding genes.

Table 1. Family of miRNAs grouped in clusters. The function of the cluster miR-17~92 is oncogenic except for miR-17.

Clusters	miRNAs in the cluster	Chromosome	Function
miR-133a-2~1-1	miR-1-1 and miR-133a-2	20	Anti-oncomiR/OncomiR
miR-1-2~133a-1	miR-1-2 and miR-133a-1	18	Anti-oncomiR/OncomiR
miR-133b~206	miR-133b and miR-206	6	Anti-oncomiR/?
miR-17~92	miR-17‾, miR-18a, miR19a, miR 20a, miR-19b-1, miR-92-1	13	OncomiR
miR-106a~363	miR-106a, miR-18b, miR-20b, miR-19b-2, miR-92-2 and miR-363	X	OncomiR
miR-106b~25	miR-106b, miR-93 and miR-25	7	OncomiR
miR-23a~27a~24-2	miR-23a, miR-27a and miR-24-2	19	OncomiR
miR-23b~27b~24-1	miR-23b, miR-27b and miR-24-1	9	Anti-oncomiR
miR-29a~29b-1	miR-29a and miR-29b-1	7	Anti-oncomiR
miR-29b-2~29c	miR-29b-2 and miR-29c	1	Anti-oncomiR
miR-34b~34c	miR-34b and miR-34c	11	Anti-oncomiR
miR-183~96~182	miR-183, miR-96 and miR-182	7	OncomiR
miR-125a~let-7c~99b	miR-125a, let-7e and miR-99b	19	OncomiR/?/AntioncomiR

Clusters	miRNAs in the cluster	Chromosome	Function
miR-let-7c~99a	let-7c and miR-99a	21	?/Anti-oncomiR
miR-100~let-7a-2	miR-100 and let-7a-2	11	Anti-oncomiR/?
miR-181a-1~181b-1	miR-181a-1 and miR-181b-1	1	OncomiR
miR-181a-2~181b-2	miR-181a-2 and miR-181b-2	9	OncomiR
miR-181c~181d	miR-181c and miR-181d	19	OncomiR
miR-199a-2~214	miR-199a-2 and miR-214	1	Anti-oncomiR
miR-302s	miR-302a, miR-302b, miR-302c, miR-302d and miR-367	4	Anti-oncomiR

2. MiRNAs families altered in cervical cancer

A gene family is a group of genes with a common phylogenetic origin and possible functional homology [24]. The members of miRNA families are grouped in clusters expressed from a same transcript. However, it is important to note that not all miRNAs from a cluster are expressed at the same time even though the cluster is generated from a single transcript. Short clusters of miRNAs usually include two or three miRNAs and the larger are from four onwards [25]. The distance between miRNAs in a cluster is approximately 1 Kb [26]. Additionally, most of the clusters have evolutionary conservation implying an important biological function. It should be noted that paralogous miRNAs are located on different chromosomes, therefore it should be expected to have differential regulation and expression [27]. Members of different miRNA families have evolved to target a diverse set of transcripts [28,29] regulating several cellular signaling pathways. It this work we point out the pathways regulated by families of miRNAs grouped in 20 clusters involved in cervical cancer.

3. Cell-signaling pathways regulated by members of miRNA families expressed in clusters.

3.1 Regulation of PI3K-AKT and MAPK axis by miR-133b~206 cluster

The cluster miR-133a-2~1-1 is formed by miR-1-1 and miR-133a-2 [30] localized on chromosome 20 having 10,536 nt of distance between them. In this sense miR-1-2 and miR-133a-1 could be expressed as a cluster since the distance between them is

3219 nt long and are localized on chromosome 18. In a similar way miR-133b and miR-206 are localized on chromosome 6 and the distance between them is 4607 nt long [31]. Interestingly, miR-1 and miR-133 from the clusters miR-133a-2~1-1 and miR1-2~133a-1 have opposite expression in cervical cancer whereas miR-1 is down-regulated miR-133 is up-regulated [32]. MiR-133b from the cluster miR-133b~206 over-expression activates AKT1, ERK1 and ERK2 phosphorylation inducing cell proliferation and colony formation in cervical cell lines by the degradation/inhibition of mRNAs and proteins of mammalian sterile 20-like kinase 2 (MST2), cell division control protein 42 homolog (CDC42) and Ras homolog gene family member A (RHOA) [33], Figure 1. Proliferation and apoptosis are regulated through Raf-1 and MST2 activation. Raf-1 is activated by Ras inducing MEK1, ERK1 and ERK2 phosphorylation [34] whilst MST2 triggers apoptosis through PUMA and caspases activation. MST2-Raf-1 interaction inhibits MST2 activation while MST2-RASSF1A interaction activates MST2. RASSF1A and Raf-1 compete for binding sites within MST2 [34,35]. Apoptosis stimulus induces MST2-RASSF1A interaction activating MST2. AKT phosphorylates MST2 inducing Raf-1 interaction releasing RASSF1A from the complex inducing MST2 inhibition [35]. Furthermore, RhoA and CDC42 inhibit ERK1/2 activation as well as AKT signaling through PTEN activation [36-38], however it should be noted that RhoA could activate the cell signaling PI3K/AKT and ERK1/2 by the activation of Plexin-B1 [39,40] Figure 5, and probably by cdc42 too, Figure 1. It seams that a cell specific pathway axis activation or inactivation response is dependent on intensity and time [38]. The AKT and MAPK cell signaling pathways have an intricate and complex regulation to induce cancer that in part could be explained by miRNA family members misregulation. MiR-133a and miR-133b differ at single 3' terminal nt and have similar expression levels in cervical cancer [32] therefore it is possible a similar type of regulation on AKT and MAPK signaling, Figure 1.

Figure 1. Regulation of MAPK, PI3K-AKT, and G2-M checkpoint by members of the family miR-125 and members of miR-let-7c~99a, miR-125a~let-7e~99b, miR-100~let-7a-2 and miR-206~133b clusters in cervical cancer. MAPK cell signaling is triggered by MEK-1 activation via TRIB2 and HOXA1 conducing to ERK1/2 phosphorylation provoking apoptosis reduction and inducing proliferation, migration, and invasion. This pathway is regulated by down-regulation of TRIB2 and HOXA1 by the family miR-99 clustered in miR-let-7c~99a, miR-125a~let-7e~99b and miR-100~let-7a-2 that is diminished in cervical cancer. MST2 inhibits Raf-1 and activates LAST2/YAP reducing Ras-MEK-ERK activation diminishing proliferation and inducing apoptosis, respectively. The oncomiR-133b from the cluster miR-133b~206 inhibits the tumor suppressor MST2. As well as CDC42 and RhoA that in turn inactivates Plexin B1 and PTEN increasing PI3K-PDK1-AKT-mTOR augmenting translation. MiR-125a from the cluster miR-125a~let-7e~99b counteracts the effect of miR-133b by the down-regulation of PIK3CD inhibiting PI3K-PDK1-AKT-mTOR signaling. The family miR-99 reduces mTOR protein expression having an opposite effect to miR-206~133b cluster. The cell cycle is arrested by DNA damage via ATM and ATR activation consecutively by the activation of Chk1/2 blocking Cdc25 hampering CDK1 stopping transition from G2 to M. ATM and ATR prevents Bora activation conducing to Aurora A, PlK1 and MDM2 inactivation stabilizing p53 activating miR-34 family transcription. Also, p53 is activated directly by Chk1/2 and ATR and ATM. In cervical cancer PlK1 is up-regulated because miR-100 from the cluster miR-100~let-7a-2 is down-regulated promoting MDM2 increase conducing to p53 decrease with a concomitant miR-34 family shrink diminishing protein translation, proliferation and apoptosis increase. In bold are the miRNAs with effect on genes with validated experimental data.

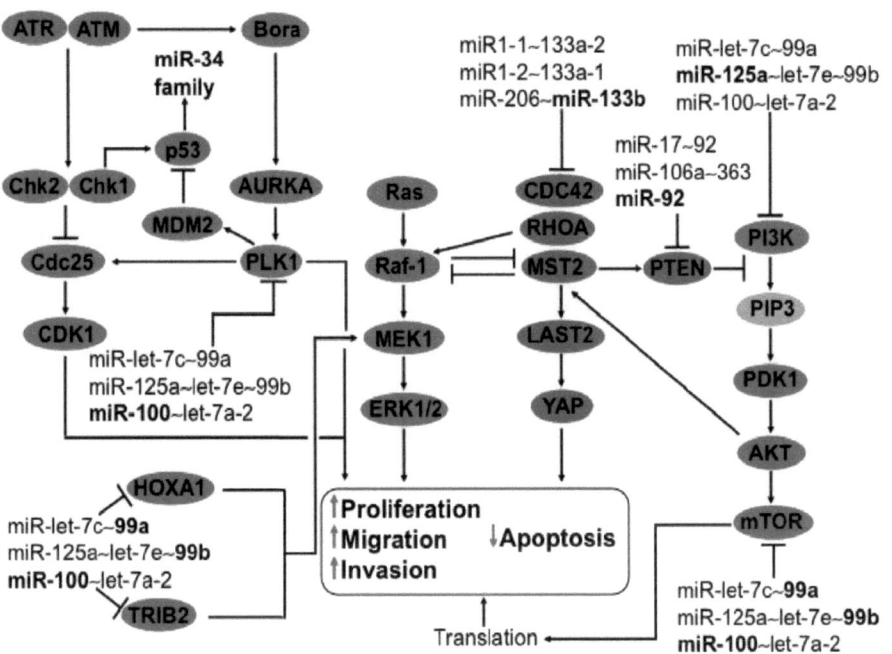

3.2 Regulation of CUL5, NOTCH, TNKS2, PTEN-PI3K-AKT axis by miR-17~92 and miR-106A~363 clusters

The families miR-17, miR-19 and miR-25 are expressed in three well-characterized clusters that have different localization on the human genome. The cluster miR17~92 is localized on chromosome 13 containing miR-17, miR-18a, miR-

19a, miR-20a, miR-19b-1, and miR-92-1. The cluster miR106a~363 is found on chromosome X and contains the miRNAs, 106a, 18b, 20b, 19b-2, 92-2, and 363. Cluster miR106b~25 is localized on chromosome 7 formed by miR-106b, miR-93 and miR-25 [31,41]. The miRNAs of these clusters are up-regulated since early steps of cervical progression, as shown in model previously proposed in a review by my group [32], except for miR-17-5p that is down-regulated and miR-18 from which there is not sufficient information in cervical cancer.

Members of the clusters miR17~92, miR-19a/b down-regulate cullin 5 (CUL5) at mRNA and protein level [42], Figure 2. CUL5 is a scaffold protein in E3 ubiquitin ligase complexes, which target substrates for ubiquitin-dependent proteasome-mediated degradation [43]. The cullin-ring ligases form a bridge with the substrate-binding cytokine-inducible suppressors of cytokine signaling (SOCS) adaptor proteins and the E2 ubiquitin-conjugating enzyme [44]. Notch transcriptional signaling substitutes SOCS for ankyrin repeat and SOCS box-containing (Asb2) from the cullin-based complex constituted by S-phase kinase-associated protein 1 (skp1), skp2, CUL5, CUL1, E47 and E2, Figure 2, inducing ubiquitination of its targets that could be involved in differentiation and proliferation [45,46]. MAPK signaling is needed for E47 degradation [47], Figure 2. A dual effect of these cell-signaling pathways are possible in cancer development, on one hand a turn-on of systems that favor tumor properties by phosphorylation and on the other hand inducing ubiquitination-degradation of tumor suppressor genes by regulation of specific proteins of ubiquitination-degradation system.

In line with this thought, miR-20a-5p, a member of the cluster miR17~92 up-regulates tankynase 2 (TNKS2) in a sequence dependent form inducing colony formation, migration, and invasion [48]. TNKS2 promotes wnt-catenin pathway liberating the inhibitory complex formed by β-catenin, glycogen synthase kinase 3 (GSK-3), adenomatous polyposis coli (APC), and AXIN1 by AXIN1 degradation inducing stabilization and freedom of β-catenin activating transcription factor/ lymphoid enhancer-binding factor 1 (TCF/LEF) [49]. Wnt-β-catenin activates Notch signaling through JAG1 ligand of Notch receptor [50] potentiating the hallmarks of cancer [51], Figure 2.

The clusters miR17~92 and miR106a~363 is involved in the positive regulation of PI3K-AKT signaling pathway. Mature miR-92 that could be formed form the cluster miR17~92 and/or miR106a~363, down-regulates PTEN in cervical carcinoma cells turning-on PI3K-AKT signaling [52], Figure 2. One of most important points in the carcinogenesis process is the transduction of cell signaling favoring the hallmarks of cancer. Interestingly, phosphatidylinositol-4,5-bisphosphate 3-kinase (PI3K) signaling is activated through Notch pathway a common oncogene in cervical cancer [53,54]. The members of the clusters miR17~92, miR106a~363 and miR106b~25 could potentially regulate CUL5, TNKS2, PTEN because they have seed sequence that hybridized with their mRNA (microRNA.org, MIRDB, Target Scan). The cluster miR17~92, miR106a~363 and miR106b~25 function as onco-miRs activating several cellular signaling inducing cervical cancer development.

Figure 2. Regulation of CUL5, JAG-NOTCH, TNKS2, and PTEN-PI3K-AKT cell signaling by members of miR-17~92 and miR-106A~363 clusters in cervical cancer. E3 ubiquitin ligase complexes formed by CUL5-E2-SOCS-JAK2/3 and CUL1-skp1-skip2-E47-E2 are activated by Abs2 and SOCS interchange potentiated through JAG1-NOTCH axis. The cell signaling GS3K-β-catenin-APC-AXIN-1-TCF/LEF is activated by AXIN-1 inhibition that triggers JAG1-NOTCH pathway that in turn accelerates the interchange of SOCS by Abs2. Additionally, CUL1-skp1-skip2-E47-E2 is activated by phosphorylation of E47 via Ras-Raf-MEK-1-ERK1/2. The miR-19a and miR-19b grouped in the clusters miR17~92 and miR106a~363 down-regulated CUL5 increasing proliferation, migration, and invasion and diminish apoptosis. MiR-20a a member of the cluster miR17~92 increases TNK2 that inhibits AXIN-1 activating TCF/LEF inducing JAG1-NOTCH1 signaling. Other member of the clusters miR17~92 and miR106a~363, miR-92, hinders PTEN expression supporting PI3K-PDK1-AKT1 signaling spreading proliferation, migration, invasion and constraining apoptosis. In bold are the miRNAs with effect on genes with validated experimental data.

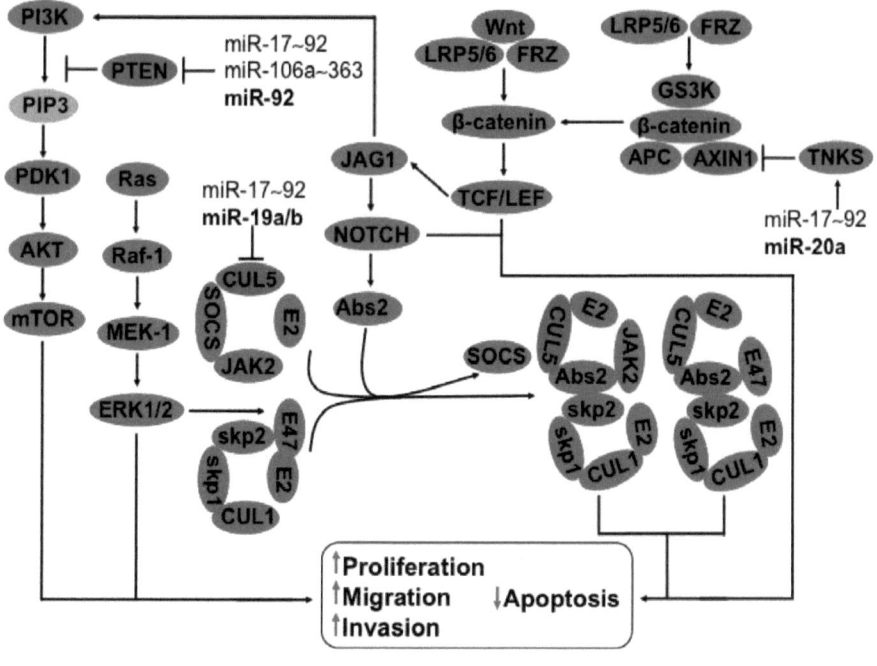

3.3 Regulation of uPA-PLG-MMP and JAG-NOTCH axis by miR-23b~27b~24-1 cluster

The families miR-23, miR-24 and miR-27 are organized in two clusters, Table 1. The cluster miR-23a~27a~24-2 is localized on chromosome 19 while 23b~27b~24-1 on chromosome 9 [25,31,55]. In cervical carcinoma cells it was shown that miR-23b diminished uPA expression via 3'-UTR mRNA binding [56], Figure 3. This protein is a serine protease that degrades plasminogen to plasmin activating metalloproteinases inducing extracellular matrix degradation. Recently it was shown an increase of Notch receptors expression triggering Notch pathway by uPA activation [57]. Additionally, JAG-Notch activation stimulates uPA transcription [58,59] inducing a mutual signaling activation exacerbated by miR-23b absence favoring cancer progression, Figure 3. The clusters miR-23a~27a~24-2 and 23b~27b~24-1 have a high similarity however it appears that the members of these clusters have opposite profile expressions whereas miR-27a is up-regulated, miR-23b and miR-27b are down-regulated in cervical cancer [32]. MiR-23a and miR-27a and miR-27b from the two clusters have sequences that may potentially regulate uPA (microRNA.org, MIRDB and Target Scan). It should be addressed if miRNAs of these clusters regulated the same targets to elucidate the participation of these clusters in cervical cancer advance.

Figure 3. Regulation of JAG1-NOTCH1-uPA, Cyclin D-CDK4/6-p21-p27, and PI3K-PDK1-AKT1 cell signaling by members of 23b~27b~24-1, miR-29a~29b-1, miR-29b-2~29c, miR17~92, miR183~96~182, and miR-302s clusters. MiR-34a, a member of miR-34 family, down-regulate the JAG1 and NOCTH1. JAG1 is activated by uPA inducing ECM reorganization via plasminogen-plasmin-MMP inducing the hallmarks of cancer. Protein expression of uPA is decreased by miR-23b a member of the cluster 23b~27b~24-1. The complex cyclin D-CDK4/6 phosphorylates RB inducing the liberation of E2F favoring c-Myc, c-Fos and COX-2 transcription provoking cellular division. Cellular division progression is regulated by the inhibition of cyclin D via miR-302s cluster and CDK4/6 through p18Ink4c, p21 and p27. PI3K-PDK1-AKT1 axis decreases p21 and p27 protein expression delivering CDK4/6 to be complexed with cyclin D. This signaling pathway is regulated by the down-regulation of AKT by miR-302s cluster increasing p21 and p27 protein inhibiting the complex cyclin D-CDK4/6. CDK4/6 is inhibited by p18Ink4c that in turn is down-regulated by miR-34a inducing apoptosis and proliferation diminish. Other point of control is given by miR-29a~29b-1 cluster hindering CDK4/6 provoking inhibition of the complex cyclin D-CDK4/6. YY-1, a target of the cluster miR-29a~29b-1, induces transcription of c-Myc in absence of this cluster that is common in cervical cancer. Cell cycle continuity is dependent on the levels and the complex formed of cyclin D-CDK4/6 and cyclin E-CDK4/6. FOXO1 could directly hinder CDK4/6 impeding the formation and activation of the complex cyclin D-CDK4/6 and cyclin E- CDK4/6 and indirectly via p27. Cyclin D is indirectly up-regulated by miR-182, one member of the cluster miR183~96~182, through FOXO1 down-regulation conducing to proliferation, migration, invasion and apoptosis induction. These cellular processes are enhanced via FOXO1 phosphorylation by PI3K-PDK1-AKT that is recognized by skp2 a subunit of skip1/cul1/F-box ubiquitin protein complex targeting to degradation via proteasome. MiR-92 diminishes the expression of PTEN triggering PI3K-PDK1-AKT signaling conducing to FOXO1 reduction. In bold are the miRNAs with effect on genes with validated experimental data.

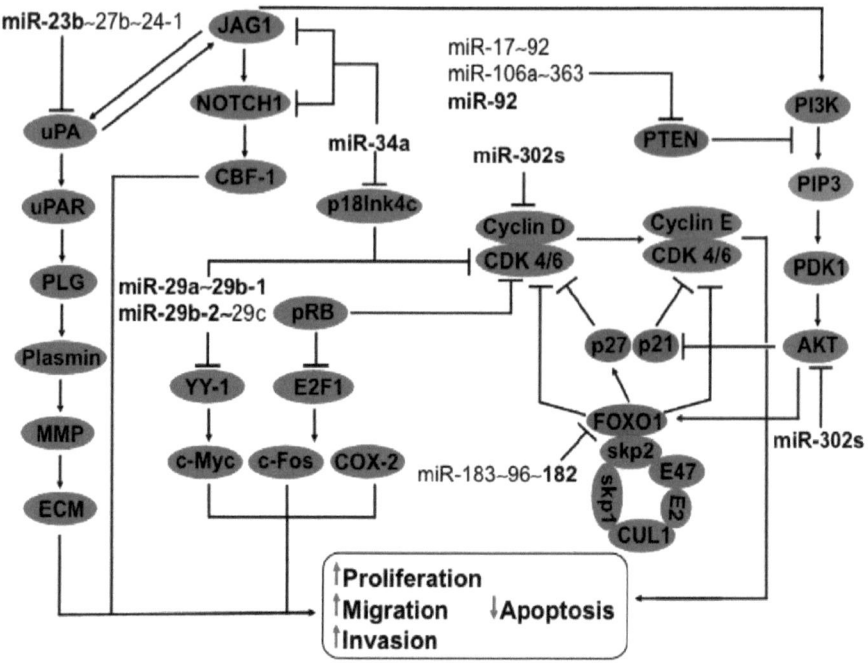

3.4 Regulation of HSP47, YY-1 and CDK6 axis by miR-29a~29b-1 and miR-29b-2~29c clusters

The members of miR-29 family are divided in two clusters, Table 1. In the first one, miR-29a and miR-29b-1 are localized on chromosome 7 whereas in the second miR-29b-2 and miR-29c reside on chromosome 1 [31]. Interestingly the members of the family miR-29 have differential expression pattern in cervical cancer while miR-29a and miR-29c are down-regulated the expression of miR-29b is not altered. The restoration of miR-29a inhibits mRNA and protein of Heat-shock protein 47 (HSP47) via 3'UTR. This protein belongs to serpin a super family of serine protease inhibitors and a molecular chaperone involved in maturation of collagen molecules [60]. The over-expression of HSP47 could be crucial in metastasis processes because structural and architectural changes are needed in extracellular matrix for carcinogenesis. In a similar way, changes in normal cell cycle are required in cancer. The proteins YY-1 and CDK6 are regulated in CIN 1 and cervical cancer, respectively, by miR-29a and miR-29b arguing a fine-tuning regulation in cervical carcinoma progress [32,61], Figure 3. Given the similarity between miR-29a, b and c, the members of the two clusters could regulate YY-1 and CDK6. It should be noted that the members of miR-29 family have different effects on cell cycle indicating target specificity [62]. YY-1 regulates several genes related to cancer like, HER2, COX-2, c-Myc, and c-Fox [63]. As well CDK-6 in complex with Cyclin D phosphorylates and inhibits the interaction of pRB with E2F1 inducing target transcription and proliferation [64], Figure 3.

3.5 Regulation of P18Ink4c and JAG-NOTCH-uPA axis by miR-34a

P18Ink4c is a CDK4/6 inhibitor that is increased in CIN 2 and carcinoma but not in normal cervix. *De novo* infection of human keratinocyte-derived raft tissue by oncogenic HPV increased p18Ink4c expression suggesting as one of the first oncogenes activated upon HPV infection. It is well known that oncogenic HPV reduces p53 expression as well as their effectors, like MiR-34a, a member of the family miR-34. The validated targets of miR-34a participate in cervical cancer progression, as is the case of p18Ink4c [65], Jagged, and Notch1 affecting cell cycle and Notch signaling directly and indirectly through uPA, respectively [66], Figure 3. MiR-34 family members share more than 80 % of sequence similarity so it is possible to have targets in common between miR-34a, and miR-34b and miR-34c which are clustered on chromosome 11 separated only by 418 nt [31]. Actually, Notch and Jag1 are potential targets of the cluster miR-34b~34c (microRNA.org, Target Scan and MIRDB). The expression of this cluster could potentiate the down-regulation of these proteins affecting Notch signaling balancing proliferation/apoptosis signals.

3.6 FOXO1-cullin-rings axis are regulated by the cluster miR183~96~182

The cluster miR183~96~182 is localized on chromosome 7 [67]. MiR-182 and miR-183 are up-regulated [68] while miR-96 is down-regulated in cervical cancer [69]. The inhibition of miR-182 diminishes tumor growth and the over-expression inhibits apoptosis and increases S phase cell cycle by down-regulating FOXO1 protein in cervical carcinoma [68]. FOXO1 is an inhibitor of cyclin/CDK complex like cyclin D1 and D2-CDK4 and cyclin E-CDK2, Figure 3. FOXO1 phosphorylation by PI3K-AKT is recognized by skp2 a subunit of skip1/cul1/F-box protein ubiquitin complex targeting it to degradation via proteasome [70], Figure 3. The function and relation of miR-182 and FOXO1 seams primordial in carcinogenesis because this axis regulated several complexes that direct cell cycle advance [71].

3.7 Regulation of PI3K-AKT and MAPK axis by miR-let-7c~99a, miR-125a~let-7e~99b and miR-100~let-7a-2 clusters

MiR-99 and miR-125 families are evolutionarily related and are clustered with members of the family let-7. Interestingly miR-125a, let-7e and miR-99b are localized on chromosome 19 separated only by 468 and 642 nt, respectively. Let-7c and miR-99a are clustered on chromosome 21 separated by 739 nt as well as miR-100 and let-7a-2 that are grouped on chromosome 11 with 5707 nt of distance among them. MiR-99a and miR-125b-2 are localized on chromosome 21 separated by 52068 nt likewise miR-100 and miR-125b-1 are on chromosome 11 with 52385 nt of distance between them. If these miRNAs are expressed as cluster is not known however they regulate similar genes [31] impacting on common cell signaling.

MiR-99a inhibits protein synthesis of tribbles pseudokinase 2 (TRIB2), Homeobox A1 (HOXA1) and mechanistic target of rapamycin (serine/threonine kinase) (mTOR) via 3'UTR [72], Figure 1. Potentially TRIB2 could be also regulated by miR-125b-2 (MIRDB and Target Scan). The expression of miR-99a and miR-125b-2 is down-regulated in cervical cancer [32] causing TRIB2, HOXA1 and mTOR over-expression. The oncogene TRIB2 is increased in several carcinomas [73,74] probably controlling the specificity of the activation of mitogen-activated protein kinases (MAPK) [75] increasing MAPK signaling, Figure 1. In similar manner the transcriptional factor HOXA1 triggers MEK1 mRNA and protein increase together with ERK1/2 phosphorylation [76] potentiating MAPK signaling enhancing the cancer hallmarks. To exacerbate the capabilities of tumor cells, mTOR pathway is constantly activated generating translational protein increase, Figure 1. Additionally, miR-99b down-regulates HOXA1 and mTOR [77]. In cervical cancer the expression of miR-99b and miR-125a are opposite [32] suggesting a specific selection of miR-125a over miR-99b resulting in HOXA1 and mTOR over-expression. MiRNAs binding capacity does not always results in protein synthesis inhibition, as is the case of miR-20a on TNSK2 regulation [48].

MiR-100 and miR-125b have similar profile expression arguing an anti-oncogenic function. MiR-100 of the cluster miR-100~let-7a-2 and miR-125b-1 are down-regulated in early steps of cervical cancer progression [32]. In these sense miR-100 inhibits PLK1 [78] a protein that participates in G2/M phase check-point regulation to block cell progression induced by DNA damage, however some tumor cells override this check-point [79]. Upon UV-DNA damage, the kinases ATM and ATR are activated inducing Chk1/2 phosphorylation and activation triggering Cdc25 phosphatase inhibition resulting in CDK1 phosphorylation and G2/M arrest. Additionally, ATM/ATR inactivates Bora leading to cascade inhibition of Aurora A,

PLK1 and Cdc25, Figure 1. The over-expression or constant activity of PLK1 conduces MDM2 activation and concomitant p53 degradation via ubiquitination. Likewise, Cdc25 activation and CDK1 dephosphorylation favors G2/M progression [80]. Interestingly, miR-100 regulates these cell-signaling pathways making it a new player in the G2/M checkpoint. The profile expression of the cluster miR100~125b-1 and the similarity between them point forward that miR-125b could participate too in this signaling. Other important targets of miR-100 are phosphatase (CTD (carboxy-terminal domain, RNA polymerase II, polypeptide A) small phosphatase-like) (CTDSPL), enzyme N-Myristoyltransferase 1 (NMT1), Transmembrane Protein 30A (TMEM30A), and chromatin remodeler SWI/SNF Related, Matrix Associated, Actin Dependent Regulator Of Chromatin, Subfamily A, Member 5 (SMARCA5), HOXA1 and mTOR. MiR-125b inhibits PI3K/AKT pathway through down-regulation of mRNA and protein PIK3CD via 3'UTR binding conducing to protein kinase A (AKT) and mTOR phosphorylation shrink inducing tumor growth volume inhibition [81], Figure 1. Down-regulation of the targets of the families and clusters of miR-99 and miR-125 hinders proliferation and migration arguing the importance of function loss of these families in cervical carcinogenesis progression. The unbalance of signaling pathways gives advantages toward development of cancer.

3.8 Regulation of PI3K-AKT and MAPK axis by miR181a-1~181b-1 and miR181a-2~181b-2 clusters

MiR-181 family is clustered on chromosomes 1, 9 and 19. MiR-181a-1 and miR-181b-1 are localized on chromosome 1 while miR-181a-2 and miR-181b-2 are annotated on chromosome 9 and miR-181c and miR-181d are located on chromosome 19 [31]. Cluster miR-181a~181b presents advantages for cervical cancer development. It has been reported recently that miR-181a confers radiochemo-resistance by diminishing mRNA and protein of PKCδ via 3'UTR binding, decreasing caspase 3/7 activity hindering apoptosis [82,83], Figure 4. MiR-181b down-regulates Adenylyl cyclase (AC) restricting cAMP production promoting cell proliferation and apoptosis diminish [84]. cAMP production conduces to PKA activation inducing transcription of smac/Diablo by CREB which leads to caspases activation [85], Figure 4. This pathway seams regulated by all members of miR-181 family clustered on distinct chromosomes in agreement with their similarity [86] and with in silico predictions (microRNA.org, MIRDB and Target Scan).

Figure 4. Regulation of AC-PKA-smac-caspase3 and PKCδ cell signaling by members of miR-181a-1~181b-1 and miR-181a-2~181b-2 clusters. Adenylyl cyclase (AC) induces PKA activation generating CREB phosphorylation that turns-on smac/Diablo causing caspase 3 activation. In addition PKCδ activates caspase 3 and caspase 7 as well. MiR-181b localized in clusters miR-181a-1~181b-1 and miR-181a-2~181b-2, down-regulates AC, as well as miR-181a inhibits PKCδ protein expression. In bold are the miRNAs with effect on genes with validated experimental data.

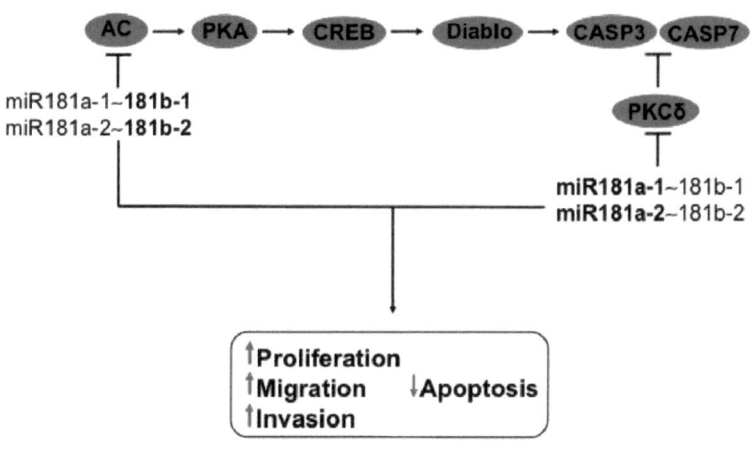

3.9 Regulation of MEK, JNK and Bcl2-l2 axis by miR-214~199a-2 cluster

Family miR-199 is formed by miR-199a-1, miR-199a-2 and miR-199b [31]. One member of family miR-199, miR-199a-2, is clustered with miR-214 [87,88] on chromosome 1 separated by 5628 nt [31], Table 1. MiR-214 functions as anti-oncomiR in cervical cancer in this respect it is not known if miR-199a-2 has the same role or it participates in cellular pathways regulated by miR-214. The targets of miR-214 are implicated on several cellular pathways. For example, the signaling through trans-membrane receptor Plexin-B1 induced cell survival, proliferation, angiogenesis, invasion, and metastasis in cervical cancer [89]. Semaphorin D4 binds to Plexin-B1 inducing the activation of RhoA that in turn activates Raf protein triggering MEK/ERK signaling [40]. Cell-signaling pathway MEK/JNK is inhibited by miR-214 targeting MEK3 and JNK1 decreasing cell proliferation [20]. Importantly these pathways have different targets. On one hand MEK3 induces p38 activation by phosphorylation resulting in MSK1 triggering. On the other hand JNK1 activates MSK2, Figure 5. Additionally, miR-214 reduces GALNT-7 protein expression affecting proliferation, migration, and invasion in cervical cell lines [90] and controls cell death through mRNA and protein down-regulation of anti-apoptotic proteins like Bcl-2l2 that in turn induces Bax increment and Caspase 9, 8, and 3, activation triggering intrinsic/extrinsic apoptosis pathways [91], Figure 5.

Figure 5. Regulation of Bcl-12l2-Bax-CASPS, MST1-MEK-4/7-JNK-1-MSK2, MST1-MEK-3-p38-MSK1, PI3K-PDK1-AKT, and Ras-Raf-MEK1-ERK1/2 cell signaling by miR-214~199a-2 cluster. The anti-oncomiR-214 from the cluster miR-214~199a-2 promotes apoptosis by down-regulation of Bcl-12-l2 permitting Bax activation with a consequent activation of caspases 9, 8 and 3. Furthermore, miR-214 inhibits JNK1 and MEK3, participants of cell signaling MST1-MEK-4/7-JNK-1-MSK2 and MST1-MEK-3-p38-MSK1, respectively. In addition miR-214 controls MAPK activation by down-regulation of Plexin-B1, that in contact with Sema 4D activates RhoA that activates Raf-1 triggering MEK-1 and ERK1/2 phosphorylation together with PI3K-PIP3-PDK1-AKT promoting the hallmarks of cancer. In bold are the miRNAs with effect on genes with validated experimental data.

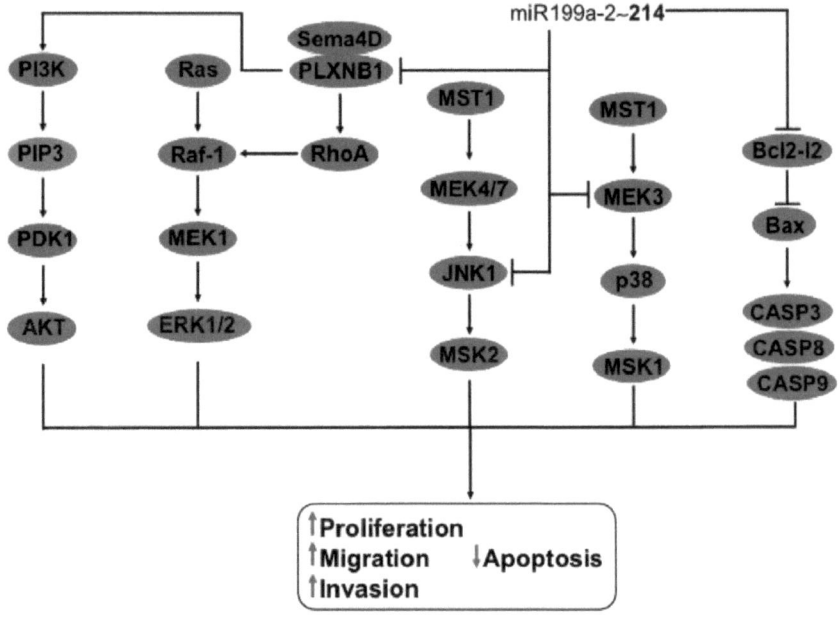

3.10 Regulation of Cyclin D and AKT axis by miR-302s cluster

The family miR-302 is formed by miR-302 a, b, c, d, e, and f. MiR-302a, miR-302b, miR-302c and miR-302d are clustered on chromosome 4 while miR-302e and miR-302f are annotated on chromosome 11 and 18, respectively [31]. Cluster miR-302s is formed by miR-302a, miR-302b, miR-302c, miR-302d, and miR-367 which down-regulates AKT and Cyclin D at protein level via 3'UTR interaction, Figure 3. Noteworthy, it was shown that AKT phosphorylation could be affected by miR-302s over-expression independently of PTEN, whose expression wasn't affected by this miRNA cluster. Furthermore, protein level and AKT phosphorylation diminish resulted in a p27 and p21 increase [92], Figure 3.

4. Altered miRNAs Expression in Cervical Carcinomas

Aberrant miRNAs expression is well recognized as a marker for several carcinomas [93]. Tumor miRNAs expression has been evaluated by numerous techniques including microarrays, sequencing, northern blotting, cloning, and reverse transcription-polymerase chain reaction (RT-PCR) [94-97]. The results of these evaluations vary between research groups. The difference could be explained by the sample preparation, the methodology used, and/or the population, however, it is possible to find expression patterns independently of the variables mentioned before if the experimented changes in miRNAs expression are strong enough and constant at least in two different works. These two characteristics in miRNAs expression could give us an insight into studying and assigning miRNAs functions in cervical cancer.

One of the first studies made in cervical cancer involving miRNAs was done in 2007. In this work, they sequenced 166 miRNAs in normal tissue, cell lines and tumor tissue. They found six miRNAs with differential expressions. Let-7b, let7-c, miR-23b, miR-143 and miR-196b were down-regulated in cell lines and tumor tissue compared with normal tissue whereas miR-21 was up-regulated [96]. Since then, a large number of studies has addressed the importance of microRNAs in cervical cancer. For example, 157 miRNAs were analyzed by RT-PCR in normal tissue and invasive squamous cervical cells resulting in 68 miRNAs over-expressed and two under-expressed in the cancer cells. Among these, the miRNAs that recorded the biggest increase were miR-9, miR-127, miR-133a, miR-133b, miR-145, miR-199a, miR-199b, miR-199s, and miR-214, while miR-149 and miR-203 showed the lowest expression [94].

A microarray analysis showed that miR-182, miR-183, and miR-210 were up-regulated and miR-128, miR-143, miR-145, and miR-195 were down-regulated in cervical carcinoma compared to normal tissue [95]. One hundred and seventy four miRNAs were cloned from normal tissue, cell lines, HPV-infected raft tissue and cervical cancer. In these experiments, miR-15b, miR-16, miR-146a, and miR-155 were over-expressed while miR-143, miR-145, and miR-128, were down-regulated [97]. Four novel miRNAs (miR-1273f, miR-1273 g, miR-5095, and miR-5096) have been discovered while searching the fragile sites related to cervical cancer. These miRNAs were noticed in SiHa, HeLa, C33-A, and tumors but not in normal tissue [98]. Another study showed that miR-886-5p was increased in cervical squamous cell carcinomas (SCC) compared to normal adjacent tissue, while miR-10a*, miR-30a*, (the star strand or passenger strand is generally degraded from miRNA duplex in miRNA biogenesis) miR-302d, miR-346, miR-518b, and miR-610 were decreased [99].

The number of miRNAs involved in cervical cancer has increased importantly. Fifteen miRNAs (miR-7, miR-18a, miR-20a, miR-20b, miR-31, miR-93, miR-141, miR-142-5p, miR-146b, miR-189, miR-200a, miR-200b, miR-210, miR-224, and miR-429) and 17 miRNAs (miR-1, miR-10b, miR-99a, miR-99b, miR-100, miR-127, miR-140, miR-143, miR-145, miR-152, miR-195, miR-214, miR-218, miR-320, miR-368, miR-376a, and miR-497) were up- and down-regulated in cervical cancer compared to normal tissue, respectively [23,100]. Eighteen miRNAs were up-regulated (miR-10b, miR-15a, miR-16, miR-17, miR-20b, miR-21, miR-93, miR-106a, miR-106b, miR-130b, miR-146b-5p, miR-155, miR-185, miR-195, miR-339-5p, miR-625, miR-941, and miR-1224-5p) and 16 were down-regulated (miR-99a, miR-100, miR-125b, miR-139-5p, miR-139-3p, miR-145, miR-199a, miR-199b-5p, miR-149, miR-328, miR-375, miR-379, miR-381, miR-497, miR-574-3p, and miR-617) as well in SCC compared to normal tissue [101].

Table 2 lists all miRNAs with evident expression changes that have been reported in at least two different studies of cervical cancer. Constant gene expression changes in cervical cancer are important in order to be able to discover genes implicated in carcinogenesis. In this sense, the increase of chromosome 5p is seen in over 50% of advanced SCC, and Drosha, a miRNA processing protein, is localized in this region. Drosha transcript levels and expression were not elevated in pre-malignant cervical squamous intraepithelial lesions contrary to malignant lesions. miRNAs most significantly associated with Drosha over-expression are implicated in carcinogenesis, suggesting that they regulate fundamental processes in cancer progression. Interestingly, they reported that let-7, miR-15b, miR-21, miR-31, miR-107, miR-125-5p, miR-191, miR-200c, miR-203, and miR-330-3p were over-expressed, whereas miR-193b was under-expressed, implying selective miRNAs expression for cancer development [102]. Additionally, it has been shown that Drosha over-expression provides invasion and migration advantages in tumor cells [103].

Table 2. miRNAs with constant change in at least two experimental studies. Star strand (miRNA*) or passenger strand is generally degraded from miRNA duplex in miRNA biogenesis.

Name of miRNA	Expression Level	Technic	Type of Tissue	Reference
miR-1	Down	Microarray	Cancer	[23]
	Down	Cloning and Sequencing	Cancer	[104]
miR-7	Up	Microarray	Cancer	[23]
	Down	RT-PCR	Cancer	[105]
	Up	Cloning and Sequencing	Cancer	[104]
miR-9	Up	RT-PCR	Cancer	[94]
	Up	RT-PCR	CIN 2, 3 and cancer	[106]
miR-10a	Up	Microarray	CIN 1, 3 and Cancer	[107]
	Up	RT-PCR	CIN 2, 3 and Cancer	[106]
	Up	RT-PCR	Cancer	[108]
miR-10b	Down	Microarray	Cancer	[23]
	Up	Microarray	Cancer	[101]
	Down	Cloning and Sequencing	Cancer	[104]
miR-15a	Up	Microarray	Cancer	[101]
	Up	Microarray and RT-PCR	Cancer	[109]
	Up	Cloning and Sequencing	Cancer	[97]
miR-15b	Up	Microarray	Cancer	[102]
	Up	Microarray	Cancer	[103]
	Up	Microarray	CIN 2, 3 and Cancer	[22]
	Up	Cloning and Sequencing	Cancer	[97]
miR-16	Down	Microarray	CIN 1, 3 and Cancer	[107]
	Up	Microarray	Cancer	[101]
	Up	RT-PCR	CIN 1, 2, 3 and Cancer	[110]
miR-17-5p	Up	Microarray	Cancer	[101]
	Down	RT-PCR	Cancer	[106]
	Down	RT-PCR	Cancer	[111]
miR-19a/b	Up	Microarray	CIN 2, 3 and Cancer	[22]
	Up	RT-PCR	Cancer	[42]
miR-20a	Up	Microarray	Cancer	[23]
	Up	RT-PCR	Cancer	[48]
	Up	Microarray and RT-PCR	Cancer	[112]
	Up	RT-PCR	Cancer	[113]

Name of miRNA	Expression Level	Technic	Type of Tissue	Reference
	Up	Microarray	Cancer	[23]
	Up	Microarray	Cancer	[101]
miR-20b	Up	Microarray and RT-PCR	Cancer	[109]
	Up	RT-PCR	CIN 2, 3 and Cancer	[106]
	Up	Cloning	Cancer	[96]
	Up	Microarrays and RT-PCR	Cancer	[102]
miR-21	Up	Northern blot and Microarray	Cancer	[114]
	Up	Microarray	Cancer	[101]
	Up	Microarray	CIN 2, 3 and Cancer	[22]
	Down	Cloning	Cancer	[96]
miR-23b	Down	RT-PCR	Cancer	[56]
	Down	Microarray	CIN 2, 3 and Cancer	[22]
	Down	Microarray	CIN 1, 3 and Cancer	[107]
miR-26a	Down	Microarray and RT-PCR	Cancer	[112]
	Down	Microarray	CIN 2, 3 and Cancer	[22]
	Down	Microarray and RT-PCR	CIN 1, 3 and Cancer	[107]
miR-27a	Up	Microarray	CIN 2, 3 and Cancer	[22]
	Up	RT-PCR	CIN 1, 2, 3 and Cancer	[110]
miR-27b	Down	Microarray	CIN 2, 3 and Cancer	[22]
	Down	Microarray	Cancer	[112]
	Down	Microarray and RT-PCR	CIN 2, 3 and Cancer	[61]
	Down	Microarray	CIN 1, 3 and Cancer	[107]
miR-29a	Down	RT-PCR	CIN 1, 2, 3 and Cancer	[110]
	Up	Microarray	CIN 2, 3 and Cancer	[22]

Name of miRNA	Expression Level	Technic	Type of Tissue	Reference
miR-31	Up	Northern blot and Microarray	Cancer	[114]
	Up	Microarray and RT-PCR	Cancer	[102]
	Up	Microarray and Northern Blot	Cancer	[103]
	Up	Microarray	Cancer	[23]
	Up	Cloning and Sequencing	Cancer	[104]
	Down	Northern Blot	Cancer	[21]
	Down	Microarray and RT-PCR	Cancer	[112]
miR-34a	Up	Microarray	CIN 2, 3 and Cancer	[22]
	Down	RT-PCR	CIN 1, 2, 3 and Cancer	[115]
miR-92a	Up	Microarray and RT-PCR	CIN 2, 3 and Cancer	[61]
	Up	Microarray	CIN 2, 3 and Cancer	[22]
	Up	RT-PCR	CIN 1, 2, 3 and Cancer	[110]
miR-93	Up	Microarray	Cancer	[23]
	Up	Microarray	Cancer	[101]
	Up	Microarray	CIN 2, 3 and Cancer	[22]
	Up	RT-PCR	Cancer	[100]
miR-99a	Down	Microarrays	CIN 1, 3 and Cancer	[107]
	Down	Microarray and RT-PCR	CIN 2, 3 and Cancer	[61]
	Down	Microarray	Cancer	[23]
	Down	Microarray	Cancer	[101]
miR-99b	Down	Microarray	CIN 2, 3 and Cancer	[22]
	Down	Microarray	Cancer	[23]
	Down	Cloning and Sequencing	Cancer	[104]
miR-100	Down	RT-PCR	CIN 1, 2, 3 and Cancer	[78]
	Down	Microarray	Cancer	[101]
	Down	Microarray	Cancer	[23]
	Down	Microarray	CIN 2, 3 and Cancer	[22]
	Down	RT-PCR	CIN 1, 2, 3 and Cancer	[110]
miR-106b	Up	Microarray	Cancer	[101]
	Up	Microarray	CIN 2,3 and Cancer	[22]
	Up	Microarray and RT-PCR	Cancer	[109]

Name of miRNA	Expression Level	Technic	Type of Tissue	Reference
	Up	Microarray	Cancer	[103]
miR-125a-5p	Up	Microarray	CIN 2, 3 and Cancer	[22]
	Down	Microarray and RT-PCR	Cancer	[112]
	Down	Microarray	Cancer	[101]
	Down	Microarray	CIN 2, 3 and Cancer	[22]
miR-125b	Down	Microarray and RT-PCR	Cancer	[112]
	Down	Cloning and Sequencing	Cancer	[104]
	Up	RT-PCR	Cancer	[94]
miR-133a	Up	Microarray, *in situ* Hybridization and RT-PCR	CIN 2, 3 and Cancer	[33]
	Up	RT-PCR	Cancer	[94]
miR-133b	Up	Microarray, *in situ* Hybridization and RT-PCR	CIN 2, 3 and Cancer	[33]
	Down	Cloning and Northern Blot	Cancer	[97]
	Down	Microarrays, RT-PCR and Northern Blot	CIN 3 and Cancer	[95]
	Down	Microarrays	Cancer	[23]
	Down	Microarray and RT-PCR	Cancer	[112]
miR-143	Down	Microarray and RT-PCR	Cancer	[116]
	Down	Cloning and Sequencing	Cancer	[104]
	Down	Microarray	CIN 1, 3 and cancer	[107]
	Up	Microarray	CIN 2, 3 and Cancer	[22]
	Up	RT-PCR	Cancer	[94]
	Down	Microarray and Northern Blot	CIN 3 and Cancer	[95]
	Down	Cloning, Microarray and Northern Blot	Cancer	[97]
	Down	Microarray	Cancer	[101]
miR-145	Down	Microarray	Cancer	[23]
	Down	Microarray and RT-PCR	Cancer	[112]
	Down	RT-PCR	Cancer	[117]
	Down	Microarray	CIN 1, 3 and cancer	[107]
	Down	Microarray	CIN 2, 3 and Cancer	[22]

Name of miRNA	Expression Level	Technic	Type of Tissue	Reference
	Up	Cloning and Northern Blot	Cancer	[97]
miR-146a	Up	Microarray	CIN 2, 3 and Cancer	[22]
	Up	RT-PCR	Cancer	[118]
miR-146b-5p	Up	Microarray	Cancer	[23]
	Up	Microarray	Cancer	[101]
	Up	Cloning	Cancer	[97]
	Up	Microarray and RT-PCR	CIN 2, 3 and Cancer	[61]
miR-155	Up	Microarray	Cancer	[101]
	Up	Microarray	CIN 2, 3 and Cancer	[22]
	Up	Cloning and Sequencing	Cancer	[104]
	Down	Northern Blot and Microarray	Cancer	[114]
miR-191	Up	Microarray	Cancer	[103]
	Up	Microarray	CIN 2, 3 and Cancer	[22]
	Down	RT-PCR and Microarray	Cancer	[102]
miR-193b	Down	RT-PCR	CIN 2, 3 and Cancer	[106]
	Up	RT-PCR, Microarray and Northern Blot	CIN 3 and Cancer	[95]
	Down	RT-PCR, Microarray and Northern Blot	CIN 3 and Cancer	[95]
miR-195	Down	Microarray and RT-PCR	CIN 2, 3 and Cancer	[61]
	Down	Microarray	Cancer	[101]
	Down	Microarray	Cancer	[23]
	Down	Microarray	CIN 2, 3 and Cancer	[22]
miR-196b	Down	RT-PCR	Cancer	[119]
	Down	Cloning	Cancer	[96]
	Up	RT-PCR	Cancer	[94]
	Down	Microarray	CIN 1, 3 and cancer	[107]
miR-199a	Down	Microarray	Cancer	[101]
	Down	Microarray	CIN 2, 3 and Cancer	[22]
miR-200a	Up	Microarray	Cancer	[23]
	Up	RT-PCR	Cancer	[100]
miR-200a*	Up	Microarray	CIN 2, 3 and Cancer	[22]
	Up	Cloning and Sequencing	Cancer	[104]

Name of miRNA	Expression Level	Technic	Type of Tissue	Reference
	Up	Microarray	Cancer	[103]
miR-200c	Up	Microarray	CIN 2, 3 and Cancer	[22]
	Down	RT-PCR	Cancer	[112]
	Up	Microarray	Cancer	[23]
	Down	RT-PCR	Cancer	[97]
	Up	Microarray	Cancer	[102]
	Down	Microarrays	CIN 1, 3 and cancer	[107]
miR-203	Down	Microarray	CIN 2, 3 and Cancer	[22]
	Down	RT-PCR	Cancer	[120]
	Down	RT-PCR	Cancer	[113]
	Down	RT-PCR	CIN 2, 3 and Cancer	[106]
	Up	Cloning and Sequencing	Cancer	[104]
	Down	RT-PCR	Cancer	[94]
	Down	Microarray	CIN 1, 3 and cancer	[107]
miR-205	Up	RT-PCR	Cancer	[121]
	Up	RT-PCR	Cancer	[104]
	Up	RT-PCR, Microarray and Northern Blot	CIN 3 and Cancer	[95]
	Up	RT-PCR, Microarray and Northern Blot	CIN 3 and Cancer	[95]
miR-210	Up	Microarray	Cancer	[23]
	Down	Microarray	CIN 2, 3 and Cancer	[22]
	Down	Northern Blot and Microarray	Cancer	[114]
miR-214	Down	Microarray	Cancer	[23]
	Down	RT-PCR	Cancer	[91]
	Up	RT-PCR	Cancer	[94]
	Down	RT-PCR, Microarray and Northern Blot	CIN 3 and Cancer	[95]
	Down	Microarray	Cancer	[23]
	Down	RT-PCR	Cancer	[122]
miR-218	Down	Microarray	CIN 2, 3 and Cancer	[22]
	Down	RT-PCR	Cancer	[123]
	Down	RT-PCR	CIN 1, 2, 3 and Cancer	[124]
miR-224	Up	Microarray	Cancer	[23]
	Up	RT-PCR	Cancer	[125]

Name of miRNA	Expression Level	Technic	Type of Tissue	Reference
	Down	Microarray	Cancer	[101]
	Down	Microarrays and RT-PCR	CIN 2, 3 and Cancer	[61]
miR-375	Down	Microarray	CIN 2, 3 and Cancer	[22]
	Down	RT-PCR	CIN 2, 3 and Cancer	[126]
	Down	Microarray and RT-PCR	Cancer	[112]
	Down	RT-PCR	CIN 1, 2, 3 and Cancer	[127]
miR-424	Down	RT-PCR	Cancer	[128]
	Down	RT-PCR	Cancer	[106]
	Down	Microarray	Cancer	[23]
miR-497	Down	Microarray	Cancer	[101]
	Down	Microarray	CIN 2, 3 and Cancer	[22]

5. miRNAs Implicated in Cervical Cancer Progression

miRNA expression profiles have shown progressive expression changes between normal, cervical intraepithelial neoplasia (CIN) 1, 2, 3, and SCC. In a recent work, miR-29a, miR-99a, miR-195, and miR-375 were shown to be down-regulated in HPV16 CIN 2 and 3 *versus* normal tissue, and the expression continued diminishing in SCC. In contrast, miR-92a and miR-155 had an opposite expression pattern in CIN 2, 3 and SCC [61]. Later a constant and progressive reduction of miR-29a and miR-100 from CIN 1 to CIN 3 and cervical cancer was shown, while miR-16, miR-25, miR-27a, miR-92a, and miR-378 recorded an increased expression [110]. However, miR-375 participation in progression is not clear because its expression was decreased in CIN 2 and 3 compared to SCC but not between normal tissue and CIN 2 and 3, suggesting a participation in the latter stages of cancer development [126]. miR-375 expression in cervical cancer progression needs further work to elucidate its clear participation in carcinogenesis.

Other works have shown differential expression profiles between normal cervical tissue, CIN, and cervical cancer. A progressive expression reduction of miR-143, miR-145, and miR-218 was shown in CIN 3 toward cervical cancer [95,107]. In cervical cancer and HPV infected raft tissue from pre-neoplasic lesions *versus* non-infected raft tissue, an increased expression was found in miR-15, miR-146, and miR-424, while a down-regulation was seen for miR-143 and miR-145 [97].

A different study revealed that miR-26a, miR-29a, miR-99a, miR-199a, miR-203, and miR-513 were decreased in CIN 2, 3 and carcinoma while miR-10a, miR-132, miR-148a, miR-196a, miR-302b, miR-512-3p, and miR-522 were increased [107]. The analysis between SCCs, CIN 2, 3, and normal tissue showed 33 miRNAs with concordant differential expressions. Eighteen miRNAs were up-regulated (let-7i, miR-19b, miR-21, miR-25, miR-28-5p, miR-30e, miR-34a, miR-34b*, miR-92a, miR-92b, miR-106b, miR-146a, miR-181d, miR-200a*, miR-206, miR-338-5p, miR-592, and miR-595) and 15 miRNAs were down-regulated (miR-23b, miR-134, miR-149, miR-193b, miR-203, miR-210, miR-296-5p, miR-365, miR-370, miR-493, miR-572, miR-575, miR-617, miR-622, and miR-638) [22][22]. A progressive miR-129a-5p down-regulation from CIN 1 to 3 and cervical cancer *versus* normal tissue was recently shown [129].

Until now there are some issues regarding miRNAs profile changes in different studies. For example, while the Lee group [94] showed an increase, the Yang group [20] showed a reduction in miR-214 expression. An additional study reported a miR-214 expression increase [23]. MiRNA profile changes should be taken with care because in some studies they are up-regulated and in others they are down-regulated (Table 2). It is also important to mention the unique gene expression of each patient.

6. miRNAs Regulated by HPV Oncoproteins

MicroRNAs misregulation in cervical cancer partly follows loss or gain of miRNA function after HPV integration [130,131]. Viral integration occurs mostly in transcriptionally active regions that include intron and/or exon sequences [132], affecting coding and non-coding genes. The studies so far have shown only a few miRNAs directly regulated by HPV oncoproteins. HPV L2 proteins are associated with a reduction of miR-125b expression and an increase of HPV DNA content. Recovering or reducing miR-125b expression, decreases or increases HPV DNA content, respectively [133]. L2 protein has not been classified as an oncoprotein but this new function could be the first evidence to rethink the classification of L2. MiR-125b levels could be crucial in people with HPV infection. Furthermore, it has been shown that miR-125b expression decreases in cervical cancer *versus* normal tissue [22,101]. The mechanisms leading to miR-125b reduction could be the result of DNA and histone methylation silencing [134] probably induced by HPV oncoproteins. Cells expressing HPV-16 E5 protein augment miR-146a expression while miR-324-5p and miR-203 diminish [135]. MiR-146a expression has been addressed as being important in cervical pathogenesis since it increases cell proliferation [97]. Single nucleotide polymorphism (SNPs) of miR-146a G allele significantly increases SCC risk [136,137] as the SNP alters miR-146a maturation [138]. MiR-218 is reduced by HR-HPV-E6-dependent expression [95]. Interestingly, miR-218 levels in patients with HR-HPV infection were lower than in those infected with low-risk or intermediate-risk HPV or in those who were HPV-free. Additionally, miR-218 levels were lower in high grade CIN than in those with low grade CIN [124]. Furthermore, the low serum levels of miR-218 in cervical cancer correlated with later stages of cervical cancer, adenocarcinoma, and lymphatic node metastasis [122]. Therefore, it appears that miR-218 down-regulation is involved in pathogenesis and progression of cervical cancer [124].

The actions of p53 are essential and more complex than it was thought, because it represses and/or activates coding and non-coding genes at the transcription level and during their biogenesis. P53 binds to several miRNA promoters to induce repression. Examples of this regulation are the repressions of miR-106b, miR-93, miR-25, and miR-17-5p, miR-18a, miR-19a, miR-20a, miR-19b-1, miR-92-1, and miR-106a, miR-18b, miR-20b, miR-19-2, and miR-92-2 [139,140].

HPV E6-expressing cells show a p53 null phenotype; in this context one can expect that all miRNAs regulated via p53 are going to be affected by E6. It has recently been reported that miR-34a is regulated by E6-dependent expression in cervical cancer [21]. Actually, it has been shown that pri-miR-34a decreases gradually according to cervical cancer progression. Additionally, a reduction was evident in CIN 1 compared to CIN 2 as it was also observed for CIN 2 *versus* CIN 3. It is noteworthy that miR-34a expression was lower in normal HPV positive than in normal HPV negative epithelium, showing an E6-dependent expression via p53 [115].

Remarkably, it was shown consensus sites for p53 in miR-23b promoter recording reduced levels of miR-23b in an E6 expression environment and uPA, one of its targets, shows an increased expression [56]. P53 expression seems to be determinant in cervical cancer progression, as it has been demonstrated for the transcriptional regulated miRNAs that could be activated by different signaling factors. MiR-34c expression is achieved by the p38 MAPK/MK2 pathway in p53-deficient cells [141].

On the other hand, the HPV-E7 protein affects the transcriptional function of E2F, thus it could be expected that this interference might affect miRNA expression. Binding sites for E2F1 and E2F3 were identified in the promoter of miR-15b. E2F induces transcription of miR-17-92, let-7a, let7-I, miR-15/16-2 and miR-106b [142-144]. MiR-15b expression was revealed to be highly correlated with the selected cell cycle E2F-induced genes CCNA2 (cyclin A2), CCNB1 (cyclin B1), MSH6 (mutS Homolog 6), and MCM7 (minichromosome maintenance complex component 7). HPV-E7 knockdown cell lines decrease miR-15b, MCM7, CCNB1 and CCNA2 expression, thereby inducing G1 arrest [145]. Another miRNA regulated by HPV is miR-203. It was recently reported that miR-203 is down-regulated in an HPV16-E7 dependent fashion. Furthermore, p63, a transcriptional regulator involved in carcinogenesis, is targeted by miR-203 via 3'-UTR, inhibiting protein production [146]. HPV oncoproteins alter the fine-tuning gene expression that regulates several cellular processes. Understanding this complex system will reveal the steps of the progression to cancer and suitable ways to block its advance.

7. Construction of a Multistep Model of Carcinogenesis by Expression of miRNAs and Their Targets

Based on the current research reported regarding the expression of miRNAs and their targets in cervical neoplasia's progression to cancer, we propose a multistep model of carcinogenesis in cervical cancer composed of four steps: (1) the changes that a normal tissue suffers to be transformed into CIN 1; (2) the consecutive changes suffered in CIN 1 to achieve CIN 2; (3) the sequential changes acquired during CIN 2 toward CIN 3; and finally (4) the changes occurring in CIN 3 to reach cervical cancer. Although the information available up to now is limited, we are certain of the upcoming of new evidence and we expect that new knowledge can be added to the model proposed here to better understand the process of cervical carcinogenesis.

7.1. MicroRNAs Misregulated in Step 1

Even though the majority of miRNAs misregulated in cervical tissue have been found in later stages of cancer, some microRNA expression profiles have suggested their participation in cervical pathogenesis. The miRNAs down-regulation in CIN 1 reported up to now is due to miR-26a, miR-29a, miR-34a, miR-99a, miR-100, miR-143, miR-145, miR-199, miR-203, and miR-218 (Figure 6a). Most of them have already experimentally validated targets, except for miR-26a, miR-199a, and miR-203 (Figure 6b).

Figure 6. (**a**) MicroRNAs (miRNAs) implicated in cervical cancer progression. miRNAs are accommodated in steps 1, 2, 3 and 4 according to their expression. In the upper scheme, the arrow represents progressive miRNAs reduction. In the bottom scheme, the arrow represents progressive miRNAs increase; (**b**) miRNAs and their target coding genes implicated in cervical cancer progression. In the neoplasic stage, step 1, one oncomiR is up-regulated and its target is down-regulated. Additionally, in this step, seven anti-oncomiRs are down-regulated and twenty-three of their targets are up-regulated. In step 2, five additional oncomiRs are up-regulated and twelve of their targets are down-regulated. Likewise, two additional anti-oncomiRs are down-regulated and their targets are up-regulated. In step 3, only one additional anti-oncomiR is down-regulated and its target is up-regulated. In Step 4, two additional onco-miRs are up-regulated with one target down-regulated and the other one up-regulated. In the same step, five additional anti-oncomiRs are down-regulated and ten of their targets are up-regulated, including CDK6 miR-29a target.

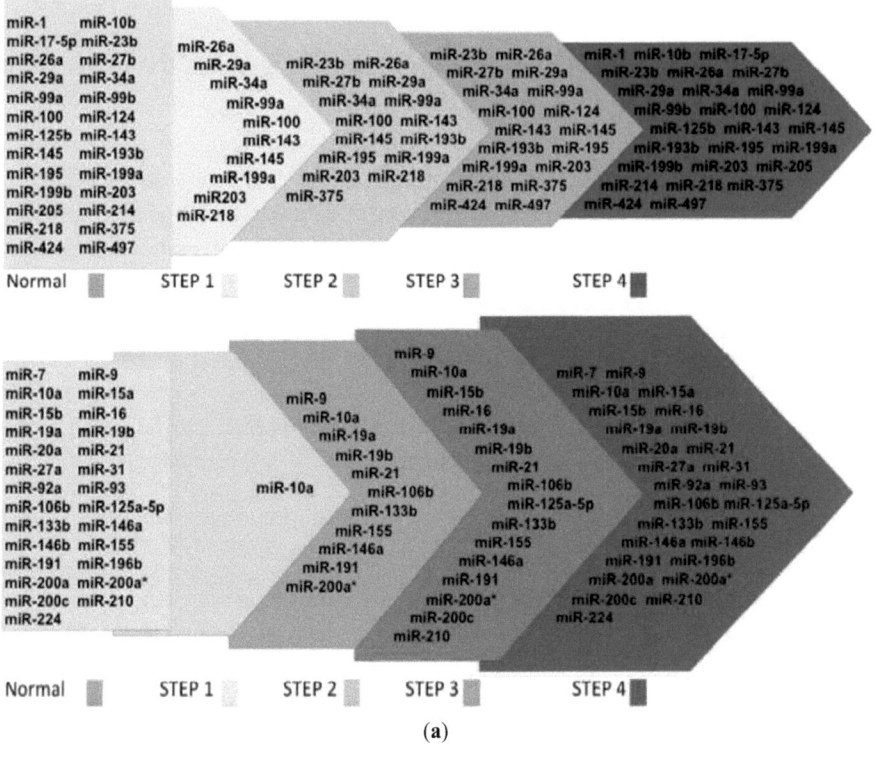

(a)

Figure 6. *Cont.*

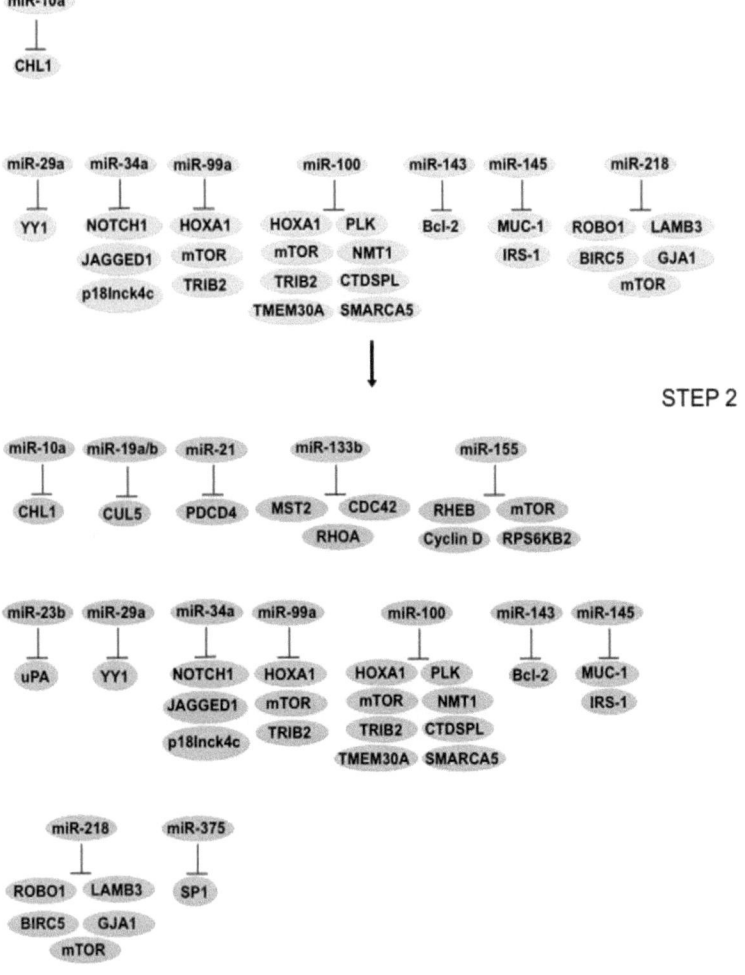

Figure 6. *Cont.*

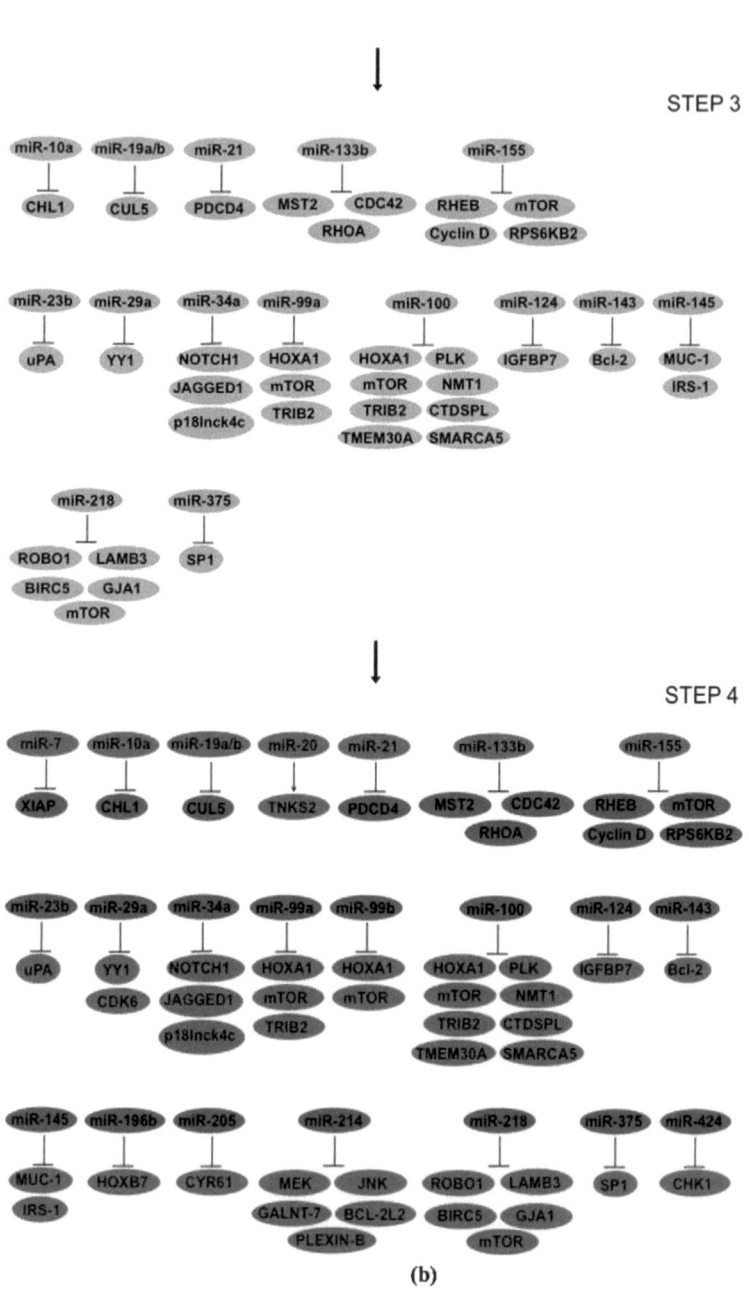

(b)

MiR-29a expression has been shown to be diminished during cervical pathogenesis, leading to apoptosis insensibility and uncontrolled cell cycles with increased Ying Yang (YY1) and CDK6 protein expression [61]. YY1 is an important transcription factor that inhibits apoptosis, and it has been shown to be over-expressed in cervical carcinomas [147]. On the other hand, CDK6 is a kinase that phosphorylates pRb releasing the transcriptional factor E2F [148]. It was shown that YY-1 protein expression began to increase from CIN 1 and 2, in contrast to CDK6 that was increased until SCC, Figure 6b. The over-expression of YY1 and CDK6 proteins inversely correlated with miR-29 expression in CIN 1, 2 and cervical cancer, respectively [61].

Another miRNA down-regulated in cervical cancer and HPV infected tissue is one of the members of miR-34 family, miR-34a. MiR-34a expression reduction is dependent on p53 regulation, therefore, expression reduction of other members of the miR34 family could also be attributable to p53 status, nevertheless, this has not been evaluated in cervical cancer. The pri-miR-34a shows a progressive reduction in CIN 1, 2, 3, and cervical carcinoma compared to normal tissue and shows an E6-dependent expression via p53 [115]. On the other hand, P18Ink4c is a CDK4/6 inhibitor that is increased in CIN 2 and carcinoma but not in normal cervix. Its participation in cancer progression seems essential because *de novo* infection of human keratinocyte-derived raft tissue by oncogenic HPV, increased p18Ink4c expression [65] contrary to miR-34a recording a reduction in cancer [21]. MiR-34a down-regulates p18Ink4c via 5'-UTR [65]. Additionally, miR-34a inhibits Jagged and Notch1 protein expression through 3'-UTR, affecting notch signaling [66]. Notch functions as a transcriptional factor implicated in controlling the expression of downstream genes associated with differentiation, cell fate specification, proliferation, apoptosis, adhesion, and angiogenesis [149]. A study showed that over-expression of the intracellular domain of Notch abolished miR-34a effect. Furthermore, urokinase-type plasminogen activator (uPA) is a serine protease that degrades extracellular matrix regulated through Notch signaling via miR-34a [66] and is considered to be involved directly in invasiveness and metastasis of cervical cell lines and probably participates in cervical cancer progression.

As well as tribbles pseudokinase 2 (TRIB2) that belongs to a family that controls the specificity of the activation of mitogen-activated protein kinases (MAPK) [75], and its increased expression has been reported in several carcinomas [73,74], suggesting an oncogenic function. One member of the family miR-99, miR-99a, inhibits protein synthesis of TRIB2 via 3'-UTR [72]. This family is composed of miR-99a, miR-99b, and miR-100 [77]. The latter has shown a gradual reduction of expression in CIN 1, 2, 3 and cervical cancer relative to normal tissue. PLK1, a key mitotic checkpoint regulatory protein usually highly expressed in cervical cancer [150] is regulated at protein but not at mRNA level by miR-100 [78]. Even though PLK1 participates in G2/M phase check-point regulation to block cell progression induced by DNA damage, some tumor cells seem to override this check-point [79]. Until now, it is not known if this failure is a consequence of miR-100 reduction. MiR-99 family is an important issue in cervical carcinoma, as it has been shown for the three members that regulate mRNA and protein levels via 3'-UTR of the transcriptional factor homeobox A1 (HOXA1) and mTOR in HaCaT cells. Also, the phosphatase (CTD (carboxy-terminal domain, RNA polymerase II, polypeptide A) small phosphatase-like) (CTDSPL), enzyme *N*-myristoyltransferase 1 (NMT1), transmembrane protein 30A (TMEM30A), and chromatin remodeler SWI/SNF related, matrix associated, actin dependent regulator of chromatin, subfamily a, member 5 (SMARCA5) are targets of miR-100. The inhibition of these genes with siRNAs constrains proliferation, as also was achieved by over-expression of miR-100. In a similar way mTOR, HOXA1, CTDSPL, TMEM30A, and SMARCA5 siRNAs hinder HaCaT cell migration contrary to NMT1 siRNA experiments [77] that show specificity of migration capability. These results permit to conclude that miR-100 aberrant expression is important in cervical carcinogenesis.

In a similar manner, miR-143/145 clusters have tumor suppressive functions because they regulate numerous recognized oncogenes showing a diminished expression in CIN 1, 2, 3, and cervical cancer [95,97,107]. An inverse expression correlation of miR-143 and Bcl-2 is well documented in cervical cancer progression arguing an important issue in cancer development [107,151,152]. MiR-143 inhibits Bcl-2 mRNA and protein via 3'-UTR, abrogating proliferation and inducing apoptosis in HeLa cells as well as volume and tumor weight reduction in nude mice [116]. Additionally, it has been shown that IRS-1 and MUC-1 negatively regulate p53 [153,154] and miR-145, an effector of p53, inhibits IRS-1 and MUC-1 inhibiting migration and invasion but not cell proliferation. As it was mentioned before, E6 expression reduces p53 levels, causing miR-145 decrease [155].

In other studies, a progressive reduction of miR-218 levels in CIN 1, 2, 3 and cervical cancer has been shown [95,122,124]. Aberrant expression of miR-218 was more prominent in high-risk than intermediate-risk and low-risk HPV and HPV-free tissue [124]. Additionally, low serum levels of miR-218 in cervical cancer correlated with later stages, adenocarcinoma, and lymphatic node metastasis [122]. In addition, it was shown that pri-miR-218 SNPs are associated with cervical cancer risk, augmenting the potential of this microRNA as a good predictive and diagnostic marker [156]. The mRNAs modulated by miR-218 are baculoviral inhibitor of apoptosis repeat containing 5 (BIRC5), roundabout (ROBO1), connexin 43 (GJA1) and laminin 5 β3 (LAMB3) [95,123,157]. BIRC5 is a cancer-specific protein that has been shown to be up-regulated in cervical cancer and participates in apoptosis, proliferation, and angiogenesis [158,159]. ROBO 1 is repressed by epigenetic changes [160], however, the mRNA has been detected in SiHa cells [102]. Connexin 43 is a transmembrane protein that functions in the organization of cell-cell communication via gap junctions in multicellular organisms and is up-regulated in cancer [161,162]. Other studies have demonstrated that LAMB3, a protein preferentially expressed in the basal lamina of the epithelium was over-expressed in cervical cancer, and its expression induced migration and invasion in several cell lines [123,163,164]. MiR-218 inhibited cell migration and invasion via LAMB3 down-regulation but no effect was seen in cell proliferation [123], indicating a specific participation of these molecules in metastasis.

Rapamycin-insentive companion of mTOR (rictor) binds to mammalian target of Rapamycin (mTOR) to form the mTOR complex-2 (mTORC2). mTORC2 induces the phosphorylation of v-akt murine thymoma viral oncogene homolog (AKT) activating proliferation and cell survival. It has been shown that miR-218 counteracts proliferation and cell survival by the down-regulation of the protein rictor inhibiting phosphorylation of AKT and increasing caspase 3 and 8 activity in HeLa cells. Interestingly, Yamamoto *et al.* [123] did not observe an effect in cell proliferation in contrast to Li *et al.* [93]. The discrepancy between these two works could be based on the levels of miR-218 achieved in HeLa cells: while Li *et al.* [165] made a stable miRNA expression, Yamamoto *et al.* [123] induced a transitory expression of miR-218. Remarkably, miR-218 over-expression diminished weight and volume of tumors in nude mice, suggesting its study in human therapy [165].

In the same step of cervical cancer development, CIN 1, miR-10a is up-regulated and negatively modulates the cell adhesion molecule L1-like (CHL1). MiR-10a regulates the mRNA and protein expression via 3'-UTR of CHL1, promoting colony formation, migration, and invasion in HeLa and C-33A cells. Interestingly, miR-10a over-expression did not have any effect on cell viability of C-33A and HeLa cells showing cell process specificity. Furthermore, an increase and decrease of miR-10a and CHL1, respectively, was shown in cervical cancer tissue *versus* normal tissue [108]. MiR-10a also shows a progressive increase in CIN compared with cervical cancer [22,107]. According to the genes mistakenly expressed during CIN 1 and analyzed in this review, a group of changes is suggested that trigger a intricate restructuration of molecular and cellular events involved in check point regulation, cell signaling through AKT and MAPK, cell adhesion molecules, and epigenetic changes affecting proliferation, cell cycle, apoptosis, migration, and invasion (Figure 6b).

7.2. MicroRNAs Misregulated in Step 2

The changes achieved in CIN 1 continue to increase in CIN 2 along with 16 new unregulated miRNAs, six down-regulated and 10 up-regulated (Figure 6a). Importantly, until now not all miRNAs have targets validated experimentally in cervical cancer as can be seen in Figures 6a,b. In this sense, only the anti-oncomiRs, miR-23b, miR-203 and miR-375; and the oncomiRs, miR-19a/b, miR-21, miR-133b, and miR-155 have experimentally validated targets, see Figure 1b. MiR-23b functions like a tumor suppressor miRNA because it is often down-regulated in HPV-associated cervical cancer contrary to one of its targets, (uPA) that is detected in cervical cancer but not in normal cervical tissues. MiR-23b diminishes uPA protein expression via interaction with its mRNA 3'-UTR. P53 absence via HR-HPV16-E6 oncoprotein was found to decrease the expression of miR-23b causing a uPA expression increase because the miR-23b promoter has a consensus p53-binding site [56].

Among the ways of gene expression regulation are the epigenetic changes that influence miRNAs expression, like DNA methylation, a hallmark for transcription silencing. A study demonstrated that the miR-203 promoter shows a great methylation status, conducive to a reduced level of mature miR-203 in cervical cancer *versus* normal tissue. Interestingly, miR-203 records a reduction in CIN 2 and 3 [22,107], causing an increase of vascular endothelial growth factor α (VEGFA). Furthermore, VEGFA was inhibited by miR-203 at the protein and mRNA level via 3'-UTR binding [120]. Additionally, p63 protein is increased upon HPV16-E7 expression in cells undergoing differentiation. This regulation has been inversely correlated with miR-203 expression, showing a mitogen-activated protein kinase-protein kinase C (MAPK-PKC) signaling dependence. It is possible that HPV16-E7 decreases miR-203 expression at the transcriptional and/or biogenesis level [146]. HPV-16 E7 interacts with HDACs [166,167] suggesting the possibility of an epigenetic silencing mechanism. Studies about miRNA processing suggest that HDAC1 is involved in pri-miRNA and/or pre-miRNA biogenesis. It has been reported that HDAC1 enhances miRNA processing via deacetylation of DiGeorge critical region gene 8 (DGCR8) [168]. Transcriptional networks are very important in cancer. The transcription factor Sp1 is inhibited by miR-375 at mRNA and protein level [169]. In cervical cancer, TGF-β1 transcriptional expression via HPV-E6/E7 protein expression through Sp1 has been reported [170]. Therefore, it is possible to expect a misregulation of genes with transcriptional sites for Sp1. The participation of Sp1 was evaluated in normal cervix *versus* carcinoma cells showing an inverse correlation with miR-375 [169]. Notably, miR-375 recorded a diminished expression in CIN 2, 3 and cervical cancer compared to normal tissue [126]. The network regulated by Sp1 and miR-375 could be fundamental in cervical carcinogenesis.

On the other hand, the oncomiR, miR-19a/b was found to inhibit cullin-5 (CUL5) also termed vasopressin-activated calcium mobilizing receptor (VACM-1) mRNA and protein via 3'-UTR. CUL5 was found to function as a tumor suppressor as it diminished cell growth and invasion in cervical carcinoma [42]. CUL5 participates in E3 ubiquitin ligase complexes targeting substrates for ubiquitin-dependent proteasome-mediated degradation [171,172], and it was shown to be involved in cellular proliferation and growth [173]. One can speculate that the target proteins for degradation by CUL5 are preferably those promoting cancer hallmarks. Likewise, an inverse expression between miR-21 and programmed cell death protein 4 (PDCD4) in invasive cervical cancer was shown. The function of PDCD4 in cell growth is probably given by blocking protein translation [174]. An increased expression of PDCD4 is recorded by miR-21 inhibition demonstrating a regulation via 3'-UTR in HeLa cells [175]. It is well documented that miR-21 is over-expressed in cervical pathologic tissue [22,94,96,97], acting as an oncomiR. MiR-21 over-expression is observed from CIN 2 to cancer compared to normal tissue either with negative or positive HPV infection [176]. In fact, an increase of miR-21 in exosomes of cervicovaginal lavage was shown in patients with cervical cancer and normal HPV positive subjects *versus* HPV negative patients [118]. Additionally, it was shown that miR-133b expression increased AKT and MAPKs (ERK1 and ERK2) phosphorylation, augmenting tumorigenesis. A gradual increase of miR-133b expression and AKT, ERK1, and ERK2 phosphorylation in CIN 2, 3 and cervical carcinoma were shown. AKT and ERK signaling are modulated via miR-133b down-regulation of mammalian sterile 20-like kinase 2 (MST2), cell division control protein 42 homolog (CDC42) and Ras homolog gene family member A (RHOA) at mRNA and protein level. MiR-133b increase and MST2 decrease, augments cell proliferation and colony formation in cervical cell lines [33].

P53 miRNA effectors inhibit cellular processes by altering their target expression. Some miRNAs and targets of p53 miRNA effectors regulate p53 levels. Over-expression of miR-155 increases p53 mRNA levels and inhibits mRNA cyclin D expression inhibiting CasKi cell proliferation, migration, and invasion [177]. Moreover, miR-155 down-regulates Ras homolog enriched in brain (RHEB), RAPTOR independent companion of MTOR, complex 2 (RICTOR) and ribosomal protein S6 kinase, 70kDa, polypeptide 2 (RPS6KB2) via 3'-UTR inhibiting mTOR-AKT signaling [178]. It seems that miR-155 acts like a tumor suppressor; nevertheless, its expression has been reported to be elevated in several works, see Table 2. It would be very interesting to evaluate the effect of miR-155 in other cell lines and/or mouse models of cervical cancer besides CaSki and HeLa cells. P53 regulation by miR-155 needs further investigation to have an accurate model of miRNA-P53 and P53-miRNA regulation. In CIN 2, the genes altered are involved in Notch signaling, transcription networks, proteasome system, protein translation, MAPK and AKT cell signaling, and cell cycle signaling giving advantages toward cervical carcinogenesis.

7.3. MicroRNAs Misregulated in Step 3

The changes observed in CIN 2, in addition to 3 and 5, lead to miRNAs down- and up-regulation, respectively, and are guidelines for the cellular and molecular changes toward CIN 3, see Figure 6a. We discuss the findings of miR-124 because its target, insulin-like growth factor BP7 (IGFBP7), and epigenetic methylation status have been recorded in CIN 3. MiR-124 is down-regulated by DNA methylation, resulting in IGFBP7 increase at the mRNA and protein level [179]. IGFBP7 has been implicated in cervical cancer, and it may influence the persistence of HR-HPV infection [180,181]. Furthermore miR-124 methylation is increased with cervical cancer progression showing a higher DNA methylation in CIN 3 and cervical cancer than in HR-HPV-positive tissues. Additionally, miR-124 restoration inhibits proliferation and migration implying that miR-124 methylation provides advantages for carcinogenesis [179]. The changes achieved in CIN 3 are related to the methylated environment and growth factors that could trigger down- and/or up-regulation of non-coding and coding genes permitting the acquisition of the cancer hallmarks that finally lead to the final step, cervical cancer.

7.4. MicroRNAs Misregulated in Step 4

More than 50 misregulated miRNAs have been reported in cervical neoplasias and cancers (Table 2). According to the data analyzed in this study, we have information of 19 and 16 miRNAs that are down- and up-regulated, respectively, in CIN 1, 2, 3 toward cervical cancer (Figure 6a). Despite the absence of information on the expression of 12 miRNAs in CINs, see Figure 6a, in this Step we discuss 7 miRNAs based on the knowledge of their validated targets in cervical cancer and their respective changes. The miR-17-5p, miR-125b-5p, miR-196b, miR-205, and miR-214 function as anti-oncomiRs while miR-7 and miR-20a work as oncomiRs. The decrease or increase of anti-oncomiRs and oncomiRs discussed in the last step are important in cervical cancer, however, their importance in cancer progression is still unknown because the reports mentioned here were only analyzed in normal tissue *versus* cervical cancer tissue, however, their expression was not evaluated in intermediate lesions.

An inverse correlation between miR-17-5p and the tumor protein P53-induced nuclear protein 1 (TP53INP1) has been shown. TP53INP1 responds and binds to p53, increasing p21 expression at the transcriptional level through promoter binding [182]. Notably, TP53INP1 was shown to be over-expressed in cervical cancer compared to normal tissue. Furthermore, the ectopic expression of TP53INP1 in cervical cell lines led to an increased proliferation. Additionally, it was shown that miR-17-5p down-regulated TP53INP1 at the mRNA and protein level via 3'-UTR binding, inhibiting proliferation and inducing apoptosis in cervical cell lines [111]. P21, BAX, PIG3 and MDM2 promoters are regulated by the interaction of TP53INP1 with p53 [182], therefore it is possible that some miRNA promoters regulated by P53 are regulated by the interaction with these transcription factors as well. It was shown that P53 binds to the promoter of miR-17-5p and suppresses its expression [183], but it is unknown if TP53INP1 participates in this regulation. It is important to mention that it is not known whether TP53INP1 tumor suppressor activity is dependent on p53 status, because p53 is down-regulated through HPV oncoprotein E6 expression in cervical cancer. TP53INP1 and p53 activity in miRNA promoter's regulation needs further investigation, because transcriptional networks drive cell growth, cell division, and numerous cell signaling pathways that are regulated through the PI3K/AKT signaling pathway. This pathway has been shown to be on the one hand inhibited by miR-125b through down-regulation of mRNA and protein PIK3CD via 3'-UTR binding, leading to a decrease of protein kinase A (AKT) and mTOR phosphorylation and thereby inhibiting tumor growth and promoting apoptosis [81]. On the other hand miR-125b inhibits BAK1 protein synthesis via 3'-UTR binding mRNA, thereby inhibiting apoptosis and increasing tumor growth volume [184]. In the former study by Cui *et al.* [81], apoptosis was analyzed with a transitory miR-125b mimic in HeLa cells while in the later study by Wang *et al.* [184], the analysis was performed with transitory plasmids expressing pre-miR-125b. Another difference that could be noted was in reference to the controls. The work of Cui *et al.* [81] shows 2.2% of apoptotic cells, whereas that of Wang *et al.* [184] shows 7.5% apoptosis. Mice experiments performed by each group can not be compared because the work by Cui *et al.* [81] was done with HeLa cells expressing miR-125b, inhibiting tumor growth volume, while the work by Wang *et al.* [184] was done with HeLa cells expressing OCT4, increasing tumor growth volume. OCT4 over-expression induced the expression of numerous miRNAs such as miR-20a, miR-21, and miR-200c among others; therefore, the effect could be attributable to these miRNAs. OCT4 expression was observed in cervical cancer tissue but not in cervical cell lines. It should be noted that tissues are constituted of several types of cells other than tumorigenic. MiR-125b regulates numerous genes: oncogenes and tumor suppressor genes. A balanced inclination of oncogenes or tumor suppressor regulation by miR-125b should determine the effect. More studies are needed to elucidate the function of miR-125b, nevertheless strong evidence is presented regarding the regulation of PIK3CD and BAK1 by miR-125b.

Additional genes are important in the model such as those that participate in blood vessel formation, and those important to supply nutrients, growth factors, and oxygen to tumors. In this sense it was shown that VEGF transcription is generated by the transcriptional factor homeobox B7 (HOXB7), and that HOXB7 is regulated by miR-196b at protein and mRNA level via 3'-UTR binding [119], potentially affecting the genes regulated by these two coding genes. In the regulation displayed among miRNAs and their targets, it does not always end in a down- or up-regulation of genes, as it has been shown for miR-205 that binds to cysteine-rich, angiogenic inducer 61 (CYR61) and connective tissue growth factor (CTGF) mRNA. However, only CYR61 protein and mRNA were shown to be diminished by this miRNA. MiR-205 over-expression induces cell proliferation and migration in cervical cell lines. It was shown that miR-205 augments while CYR61 and CTGF mRNAs decrease in cervical cancer tissues, probably indicating that CTGF is regulated by other miRNA [121].

Some studies have addressed the importance of signaling proteins in cancer. For example, it was shown that the signaling through the trans-membrane receptor Plexin-B1 induced cell survival, proliferation, angiogenesis, invasion, and metastasis in cervical cancer [89]. Additionally, Plexin-B1 signaling was directly blocked by reduction of mRNA and protein via 3'-UTR binding by miR-214. Furthermore, miR-214 inhibited MEK3 and JNK1 at the mRNA and protein level, both genes that are involved in cell proliferation [20]. Another process related to cellular survival in cancer is the O-glycosylation that could be used to address tumor cells [185]. Nevertheless, the enzymes that participate in tumorigenesis are unknown. Recently, the enzyme UDP-N-Acetyl-α-D-galactosamine: Polypeptide N-acetilgalactosaminyl transferase 7 (GALNT-7) was recognized to increase proliferation, migration, and invasion in cervical cell lines recording a gene expression augment in cervical carcinoma. MiR-214 binds to the 3'-UTR of GALNT-7, inhibiting protein expression [90]. Additionally, miR-214 controls cell death through mRNA and protein down-regulation of anti-apoptotic proteins like Bcl-2l2, inducing the increment of Bax, Caspase 9, 8, and 3, triggering intrinsic/extrinsic apoptosis pathways [91].

The apoptosis trigger is dependent on the cell cycle checkpoint, and most of the carcinomas override this brake by gene modifications involved in these pathways.

This is the case for miR-424 that is reduced in high-grade cervical neoplasia and is positively correlated with poor tumor differentiation, advanced clinical stage, and lymph node metastasis. The gene cell-cycle checkpoint kinase 1 (ChK1) is target of miR-424 and its inhibition decreases matrix metalloproteinase 9 expression. Cell cycle arrest in response to DNA damage is a normal activity of ChK1, but its expression was shown to be higher in high-grade carcinoma, suggesting its involvement in the pathogenesis of cancer [128].

There are only two oncomiRs identified with validated targets, but we are sure of the increase of miRNAs in this final step. MiR-7 and the X-linked inhibitor of apoptosis protein (XIAP) show a down- and over-expression, respectively, in tumors compared with normal tissues. The inhibitor of apoptosis protein XIAP functions like an E3 ubiquitin ligase targeting proteins for degradation by proteasome. Interestingly, miR-7 inhibits XIAP protein and mRNA via 3'-UTR in HeLa and C33-A cells [36]. The general mechanism of miRNA action is the decrease of gene expression, but by unknown mechanisms they could as well increase gene expression [186]. The proteins that interact with miRNA machinery biogenesis regulate miRNA maturation [187]. MiR-20a interacts with the 3'-UTR of tankynase 2 (TNKS2) up-regulating mRNA and protein expression. TNKS2 has the advantage of sustaining constant proliferation. TNKS2 is a new member of the human telomerase-associated poly (ADP-ribose) Polymerase (PARP) family and has been shown to be over-expressed in cervical cancer. TNKS2 protein binds to telomerase-binding protein TRF1 and protects the ends of linear chromosome. Therefore, ablation of TNKS2 and miR-20a inhibits colony formation, migration, and invasion. Remarkably, TNKS2 and miR-20a are high in cancer compared with normal tissue [48]. HPV16-E7 is an oncoprotein that causes chromosome alteration, therefore it could potentiate the effect of TNKS2 in carcinogenesis. In cervical carcinoma, the altered processes are related to migration, invasion, anchorage independent growth, cell cycle, and apoptosis through p53-interaction proteins, PI3K/ AKT cell signaling, growth, and angiogenesis factors, transmembrane receptors, intrinsic and extrinsic apoptosis, cell check points, apoptosis counteracts, chromosome ends, and cell cycle inhibitors.

8. Conclusions

In this overview, we identified 20 clusters implicated in cervical cancer based on literature reviewed that regulated several cell signaling governing carcinogenesis highlighting cervical cancer associated miRNAs in CIN 1, 2, 3, and cancer as well as their targets. The cell pathways PI3K-AKT- mTOR and MST1-MEK-4/7-JNK-1-MSK2, MST1-MEK-3-p38-MSK1 and Ras-Raf-MEK-ERK are activated or inhibited by anti-oncomiRs clusters (let-7c~miR-99a, miR-125a~let-7e~99b, miR- 100~let-7a-2, miR-199a-2~214 and miR-302s) and oncomiRs (miR-133a-2~1-1, miR-1-2~133a-1, miR-133b~206, miR-17~92, miR-106a~363, and miR183~96~182). The anti-oncomiRs and oncomiRs expressed in these clusters are down-regulated and up-regulated, respectively, since early steps in cervical cancer progression. However it should be noted that miRNAs processing and stability are unique because we found exceptions like miR-17-5p and miR-1 that most be down and up-regulated upon clusters expression in cervical cancer, Table1. The complex regulation of the anti-oncomiRs (miR-34a and cluster 23b~27b~24-1) and oncomiRs clusters (miR-17~92 and miR-106a~363,) could be noted in the connection between NOTCH axis and proteasome system that are connected and modulated by PI3K-AKT-mTOR and Ras-Raf-MEK-ERK. In a similar fashion the cell cycle phases G1 and G2-M and apoptosis are balanced by anti-oncomiRs (miR-34a, miR-29a~29b-1, miR-29b-2~29c, miR-214 and miR-302s) and oncomiRs (miR-17~92, miR-106a~363, miR183~96~182, miR- 181a-1~181b-1 and miR-181a-2~181b-2). In the works we reviewed here they report mature miRNAs. In this context we are missing data regarding cluster origin and function of all miRNAs expressed in the clusters. Clusters on different chromosomes have equal transcript processing and stability levels? Knowing the provenance of miRNAs it should be taken into account for therapy development. With this in mind 8 clusters function like oncomiRs and 9 as anti-oncomiRs. 3 clusters have dual activity as anti-oncomiRs and oncomiRs.

It is important to note that some of the miRNA targets in our model have not been directly evaluated in CIN. However, we assume that targets of altered miRNA are also going to be affected in carcinogenesis steps. However, this is not necessarily true, as it has been shown for miR-29a, YY1, and CDK6. MiR-29a is down-regulated and its targets are differentially regulated, while YY-1 is up-regulated in CIN 1, CDK6 is increased until SCC. Additionally, it should be noted that the expression of miRNAs that do not meet the criteria of having similar expressions in at least two studies are not included in this model. In the model that we propose, in the first step (CIN 1), miRNAs and their targets are involved in the regulation of check points, cell signaling through AKT and MAPK, cell adhesion molecules, and epigenetic changes affecting the hallmarks of cancer. Continuing with miRNA alterations in step 2 (CIN 2), cellular changes are achieved through Notch signaling, transcription networks, proteasome system, protein translation, MAPK cell signaling, and cell cycle signaling. Following is step 3, in which the changes achieved in CIN 3 are associated with epigenetic changes, although more studies are needed to further complement this step. And finally, in the cancer stage, the altered molecular processes are related to p53-interaction proteins, PI3K/AKT cell signaling, growth and angiogenesis factors, transmembrane receptor signaling, intrinsic and extrinsic apoptosis, cell checkpoints, apoptosis resistance, proteins participating in chromosome ends, and cell cycle inhibitors affecting cellular and molecular processes in carcinogenesis.

Along all the steps in cervical carcinogenesis the cell-signaling pathway with the most miRNAs implicated is the AKT pathway. Interestingly, miR-99 family, miR-125b, and miR-218 are diminished which is in contrast with the increased expression of miR-155 and miR-133b resulting in increased AKT phosphorylation. In cervical cancer progression, the AKT signaling is turned on, showing an important role in advantage acquisition of malignant transformation. Another disregulated cell pathway is MAPK cell signaling, which is a crucial factor in cancer progression. To this respect, an increased phosphorylation of MEK3/JNK and ERK1/2 by the reduced expression of miR-214 and increase of miR-133b has been shown. Further, Notch signaling is constantly activated to induce cell survival by the down-regulation of the regulators miR-34a and miR-23b that modulate several points of the signaling cascade.

The cell process most frequently affected along neoplasia progression in this review is the cell cycle checkpoint. The G1 checkpoint is regulated through CDK6 and cyclin D expression by miR-29a and the network of miR-155-p53-miR-145 and p53-miR-34. CDK6 expression is up-regulated by the down-regulation of miR-29a, and cyclin D is regulated by miR-155, however, in several studies (Table 2), this miRNA is over-expressed, hence its function is unclear. Additionally, cyclin D is inhibited indirectly by miR-145 expression via p53-p21. On the other hand, P53 is mutated in 50% of carcinomas therefore miR-34a is down-regulated and its target p18Ink4c is increased in cervical cancer because the G1 check point is mutated most likely by the alteration of the feedback of miR-155-p53-miR-145, thereby overriding G1 checkpoint cell cycle. The G2-M checkpoint is overridden by the absence of its regulators, miR-100 and miR-424.

Another cell hallmark process of carcinogenesis is apoptosis. In this work, we discovered genes that were clearly involved in this important process. It is well known that p53 participates in G1 and G2-M checkpoints and that it can trigger apoptosis. Triggering or inhibiting apoptosis is fundamental for tumor survival. In this sense, miR-143 and miR-214 inhibit anti-apoptotic proteins, while miR-7 and miR-21 have the opposite function and inhibit apoptosis by the down-regulation of the pro-apoptotic proteins. Apoptosis resistance is achieved during the early steps, as the genes involved are deregulated.

Metastasis is the final process involved in cancer, and it is characterized by the formation of new tumors starting from the cells of the primary tumor. Tumor cells migrate with a set of different cell types to make an optimal niche to survive and grow. To this end, cell vessel formation is essential. The formation of new cell vessels is increased by the down-regulation of angiogenesis regulators miR-99 family, miR-196b, miR-203, and miR-205. miRNAs and their targets are located sequentially in this cervical cancer multistep model (Figure 6a,b). Based on this initial model the miRNAs discussed here and the clusters involved in cell signaling pathways regulation could be used to evaluate therapeutic, diagnostic, and prognostic applications in cervical cancer.

Author Contributions

J.A.L. analyzed the literature, organized, and created the original idea, A.J.G.L and L.S.S.G. analyzed literature and wrote the paper, LCQ, FMTM, HHL and RGH analyzed and organized literature.

Conflicts of Interest

The authors declare no conflict of interest.

Acknowledgments

This work is a compilation of two papers published in IJMS (Int. *J. Mol. Sci.* **2014,** *15,* 15700- 15733 and *Int. J. Mol. Sci.* **2015,** *16,* 12773-12790) and was supported by CONACyT (Grant No. 177620) and PROMEP (Grant No. 17192).

References

1. Parkin, D.M.; Bray, F.; Ferlay, J.; Pisani, P. Estimating the world cancer burden: Globocan 2000. *International journal of cancer. Journal international du cancer* **2001,** *94,* 153-156.

2. Chakrabarti, O.; Krishna, S. Molecular interactions of 'high risk' human papillomaviruses e6 and e7 oncoproteins: Implications for tumour progression. *Journal of biosciences* **2003,** *28,* 337-348.

3. DiMaio, D.; Liao, J.B. Human papillomaviruses and cervical cancer. *Advances in virus research* **2006,** *66,* 125-159.

4. Dyson, N.; Howley, P.M.; Munger, K.; Harlow, E. The human papilloma virus-16 e7 oncoprotein is able to bind to the retinoblastoma gene product. *Science* **1989,** *243,* 934-937.

5. Scheffner, M.; Werness, B.A.; Huibregtse, J.M.; Levine, A.J.; Howley, P.M. The e6 oncoprotein encoded by human papillomavirus types 16 and 18 promotes the degradation of p53. *Cell* **1990,** *63,* 1129-1136.

6. Calin, G.A.; Croce, C.M. Microrna signatures in human cancers. *Nature reviews. Cancer* **2006,** *6,* 857-866.

7.   Miller, D.M.; Blume, S.; Borst, M.; Gee, J.; Polansky, D.; Ray, R.; Rodu, B.; Shrestha, K.; Snyder, R.; Thomas, S.*, et al.* Oncogenes, malignant transformation, and modern medicine. *The American journal of the medical sciences* **1990**, *300*, 59-69.

8.   Weinberg, R.A. Tumor suppressor genes. *Science* **1991**, *254*, 1138-1146.

9.   Lim, P.K.; Bliss, S.A.; Patel, S.A.; Taborga, M.; Dave, M.A.; Gregory, L.A.; Greco, S.J.; Bryan, M.; Patel, P.S.; Rameshwar, P. Gap junction-mediated import of microrna from bone marrow stromal cells can elicit cell cycle quiescence in breast cancer cells. *Cancer research* **2011**, *71*, 1550-1560.

10.   Rechavi, O.; Erlich, Y.; Amram, H.; Flomenblit, L.; Karginov, F.V.; Goldstein, I.; Hannon, G.J.; Kloog, Y. Cell contact-dependent acquisition of cellular and viral nonautonomously encoded small rnas. *Genes & development* **2009**, *23*, 1971-1979.

11.   Dalmay, T.; Edwards, D.R. Micrornas and the hallmarks of cancer. *Oncogene* **2006**, *25*, 6170-6175.

12.   Bartel, D.P. Micrornas: Genomics, biogenesis, mechanism, and function. *Cell* **2004**, *116*, 281-297.

13.   Bartel, D.P. Micrornas: Target recognition and regulatory functions. *Cell* **2009**, *136*, 215-233.

14.   Griffiths-Jones, S.; Saini, H.K.; van Dongen, S.; Enright, A.J. Mirbase: Tools for microrna genomics. *Nucleic acids research* **2008**, *36*, D154-158.

15.   Lewis, B.P.; Burge, C.B.; Bartel, D.P. Conserved seed pairing, often flanked by adenosines, indicates that thousands of human genes are microrna targets. *Cell* **2005**, *120*, 15-20.

16.   Grimson, A.; Farh, K.K.; Johnston, W.K.; Garrett-Engele, P.; Lim, L.P.; Bartel, D.P. Microrna targeting specificity in mammals: Determinants beyond seed pairing. *Molecular cell* **2007**, *27*, 91-105.

17.   Baek, D.; Villen, J.; Shin, C.; Camargo, F.D.; Gygi, S.P.; Bartel, D.P. The impact of micrornas on protein output. *Nature* **2008**, *455*, 64-71.

18. Selbach, M.; Schwanhausser, B.; Thierfelder, N.; Fang, Z.; Khanin, R.; Rajewsky, N. Widespread changes in protein synthesis induced by micrornas. *Nature* **2008**, *455*, 58-63.

19. Lopez, J.A.; Alvarez-Salas, L.M. Differential effects of mir-34c-3p and mir-34c-5p on siha cells proliferation apoptosis, migration and invasion. *Biochemical and biophysical research communications* **2011**, *409*, 513-519.

20. Yang, Z.; Chen, S.; Luan, X.; Li, Y.; Liu, M.; Li, X.; Liu, T.; Tang, H. Microrna-214 is aberrantly expressed in cervical cancers and inhibits the growth of hela cells. *IUBMB life* **2009**, *61*, 1075-1082.

21. Wang, X.; Wang, H.K.; McCoy, J.P.; Banerjee, N.S.; Rader, J.S.; Broker, T.R.; Meyers, C.; Chow, L.T.; Zheng, Z.M. Oncogenic hpv infection interrupts the expression of tumor-suppressive mir-34a through viral oncoprotein e6. *RNA* **2009**, *15*, 637-647.

22. Wilting, S.M.; Snijders, P.J.; Verlaat, W.; Jaspers, A.; van de Wiel, M.A.; van Wieringen, W.N.; Meijer, G.A.; Kenter, G.G.; Yi, Y.; le Sage, C., *et al.* Altered microrna expression associated with chromosomal changes contributes to cervical carcinogenesis. *Oncogene* **2013**, *32*, 106-116.

23. Rao, Q.; Shen, Q.; Zhou, H.; Peng, Y.; Li, J.; Lin, Z. Aberrant microrna expression in human cervical carcinomas. *Med Oncol* **2012**, *29*, 1242-1248.

24. Demuth, J.P.; De Bie, T.; Stajich, J.E.; Cristianini, N.; Hahn, M.W. The evolution of mammalian gene families. *PloS one* **2006**, *1*, e85.

25. Yu, J.; Wang, F.; Yang, G.H.; Wang, F.L.; Ma, Y.N.; Du, Z.W.; Zhang, J.W. Human microrna clusters: Genomic organization and expression profile in leukemia cell lines. *Biochemical and biophysical research communications* **2006**, *349*, 59-68.

26. Kriegel, A.J.; Liu, Y.; Fang, Y.; Ding, X.; Liang, M. The mir-29 family: Genomics, cell biology, and relevance to renal and cardiovascular injury. *Physiological genomics* **2012**, *44*, 237-244.

27. Chhabra, R.; Dubey, R.; Saini, N. Cooperative and individualistic functions of the micrornas in the mir-23a~27a~24-2 cluster and its implication in human diseases. *Molecular cancer* **2010**, *9*, 232.

28. Tehler, D.; Hoyland-Kroghsbo, N.M.; Lund, A.H. The mir-10 microrna precursor family. *RNA biology* **2011**, *8*, 728-734.

29. Wang, W.X.; Danaher, R.J.; Miller, C.S.; Berger, J.R.; Nubia, V.G.; Wilfred, B.S.; Neltner, J.H.; Norris, C.M.; Nelson, P.T. Expression of mir-15/107 family micrornas in human tissues and cultured rat brain cells. *Genomics, proteomics & bioinformatics* **2014**, *12*, 19-30.

30. Chen, W.S.; Leung, C.M.; Pan, H.W.; Hu, L.Y.; Li, S.C.; Ho, M.R.; Tsai, K.W. Silencing of mir-1-1 and mir-133a-2 cluster expression by DNA hypermethylation in colorectal cancer. *Oncology reports* **2012**, *28*, 1069-1076.

31. Kozomara, A.; Griffiths-Jones, S. Mirbase: Annotating high confidence micrornas using deep sequencing data. *Nucleic acids research* **2014**, *42*, D68-73.

32. Granados Lopez, A.J.; Lopez, J.A. Multistep model of cervical cancer: Participation of mirnas and coding genes. *International journal of molecular sciences* **2014**, *15*, 15700-15733.

33. Qin, W.; Dong, P.; Ma, C.; Mitchelson, K.; Deng, T.; Zhang, L.; Sun, Y.; Feng, X.; Ding, Y.; Lu, X., *et al.* Microrna-133b is a key promoter of cervical carcinoma development through the activation of the erk and akt1 pathways. *Oncogene* **2012**, *31*, 4067-4075.

34. Leicht, D.T.; Balan, V.; Zhu, J.; Kaplun, A.; Bronisz, A.; Rana, A.; Tzivion, G. Mek-1 activates c-raf through a ras-independent mechanism. *Biochimica et biophysica acta* **2013**, *1833*, 976-986.

35. Romano, D.; Matallanas, D.; Weitsman, G.; Preisinger, C.; Ng, T.; Kolch, W. Proapoptotic kinase mst2 coordinates signaling crosstalk between rassf1a, raf-1, and akt. *Cancer research* **2010**, *70*, 1195-1203.

36. Li, F.; Jiang, Q.; Shi, K.J.; Luo, H.; Yang, Y.; Xu, C.M. Rhoa modulates functional and physical interaction between rock1 and erk1/2 in selenite-induced apoptosis of leukaemia cells. *Cell death & disease* **2013**, *4*, e708.

37. Li, Z.; Dong, X.; Wang, Z.; Liu, W.; Deng, N.; Ding, Y.; Tang, L.; Hla, T.; Zeng, R.; Li, L., *et al.* Regulation of pten by rho small gtpases. *Nature cell biology* **2005**, *7*, 399-404.

38. Zugasti, O.; Rul, W.; Roux, P.; Peyssonnaux, C.; Eychene, A.; Franke, T.F.; Fort, P.; Hibner, U. Raf-mek-erk cascade in anoikis is controlled by rac1 and cdc42 via akt. *Molecular and cellular biology* **2001**, *21*, 6706-6717.

39. Basile, J.R.; Gavard, J.; Gutkind, J.S. Plexin-b1 utilizes rhoa and rho kinase to promote the integrin-dependent activation of akt and erk and endothelial cell motility. *The Journal of biological chemistry* **2007**, *282*, 34888-34895.

40. Aurandt, J.; Li, W.; Guan, K.L. Semaphorin 4d activates the mapk pathway downstream of plexin-b1. *The Biochemical journal* **2006**, *394*, 459-464.

41. Tanzer, A.; Stadler, P.F. Molecular evolution of a microrna cluster. *Journal of molecular biology* **2004**, *339*, 327-335.

42. Xu, X.M.; Wang, X.B.; Chen, M.M.; Liu, T.; Li, Y.X.; Jia, W.H.; Liu, M.; Li, X.; Tang, H. Microrna-19a and -19b regulate cervical carcinoma cell proliferation and invasion by targeting cul5. *Cancer letters* **2012**, *322*, 148-158.

43. Chiba, T.; Tanaka, K. Cullin-based ubiquitin ligase and its control by nedd8-conjugating system. *Current protein & peptide science* **2004**, *5*, 177-184.

44. Petroski, M.D.; Deshaies, R.J. Mechanism of lysine 48-linked ubiquitin-chain synthesis by the cullin-ring ubiquitin-ligase complex scf-cdc34. *Cell* **2005**, *123*, 1107-1120.

45. Nie, L.; Zhao, Y.; Wu, W.; Yang, Y.Z.; Wang, H.C.; Sun, X.H. Notch-induced asb2 expression promotes protein ubiquitination by forming non-canonical e3 ligase complexes. *Cell research* **2011**, *21*, 754-769.

46. Wu, W.; Sun, X.H. A mechanism underlying notch-induced and ubiquitin-mediated jak3 degradation. *The Journal of biological chemistry* **2011**, *286*, 41153-41162.

47. Nie, L.; Xu, M.; Vladimirova, A.; Sun, X.H. Notch-induced e2a ubiquitination and degradation are controlled by map kinase activities. *The EMBO journal* **2003**, *22*, 5780-5792.

48. Kang, H.W.; Wang, F.; Wei, Q.; Zhao, Y.F.; Liu, M.; Li, X.; Tang, H. Mir-20a promotes migration and invasion by regulating tnks2 in human cervical cancer cells. *FEBS letters* **2012**, *586*, 897-904.

49. James, R.G.; Davidson, K.C.; Bosch, K.A.; Biechele, T.L.; Robin, N.C.; Taylor, R.J.; Major, M.B.; Camp, N.D.; Fowler, K.; Martins, T.J., *et al.* Wiki4, a novel inhibitor of tankyrase and wnt/ss-catenin signaling. *PloS one* **2012**, *7*, e50457.

50. Katoh, M.; Katoh, M. Notch ligand, jag1, is evolutionarily conserved target of canonical wnt signaling pathway in progenitor cells. *International journal of molecular medicine* **2006**, *17*, 681-685.

51. Hanahan, D.; Weinberg, R.A. Hallmarks of cancer: The next generation. *Cell* **2011**, *144*, 646-674.

52. Yu, Y.; Zhang, Y.; Zhang, S. Microrna-92 regulates cervical tumorigenesis and its expression is upregulated by human papillomavirus-16 e6 in cervical cancer cells. *Oncology letters* **2013**, *6*, 468-474.

53. Ayaz, F.; Osborne, B.A. Non-canonical notch signaling in cancer and immunity. *Frontiers in oncology* **2014**, *4*, 345.

54. Veeraraghavalu, K.; Subbaiah, V.K.; Srivastava, S.; Chakrabarti, O.; Syal, R.; Krishna, S. Complementation of human papillomavirus type 16 e6 and e7 by jagged1-specific notch1-phosphatidylinositol 3-kinase signaling involves pleiotropic oncogenic functions independent of cbf1;su(h);lag-1 activation. *Journal of virology* **2005**, *79*, 7889-7898.

55. Zhou, Q.; Gallagher, R.; Ufret-Vincenty, R.; Li, X.; Olson, E.N.; Wang, S. Regulation of angiogenesis and choroidal neovascularization by members of microrna-23~27~24 clusters. *Proceedings of the National Academy of Sciences of the United States of America* **2011**, *108*, 8287-8292.

56. Au Yeung, C.L.; Tsang, T.Y.; Yau, P.L.; Kwok, T.T. Human papillomavirus type 16 e6 induces cervical cancer cell migration through the p53/microrna-23b/urokinase-type plasminogen activator pathway. *Oncogene* **2011**, *30*, 2401-2410.

57. Raghu, H.; Gondi, C.S.; Dinh, D.H.; Gujrati, M.; Rao, J.S. Specific knockdown of upa/upar attenuates invasion in glioblastoma cells and xenografts by inhibition of cleavage and trafficking of notch -1 receptor. *Molecular cancer* **2011**, *10*, 130.

58. Shimizu, M.; Cohen, B.; Goldvasser, P.; Berman, H.; Virtanen, C.; Reedijk, M. Plasminogen activator upa is a direct transcriptional target of the jag1-notch receptor signaling pathway in breast cancer. *Cancer research* **2011**, *71*, 277-286.

59. Bin Hafeez, B.; Adhami, V.M.; Asim, M.; Siddiqui, I.A.; Bhat, K.M.; Zhong, W.; Saleem, M.; Din, M.; Setaluri, V.; Mukhtar, H. Targeted knockdown of notch1 inhibits invasion of human prostate cancer cells concomitant with inhibition of matrix metalloproteinase-9 and urokinase plasminogen activator. *Clinical cancer research : an official journal of the American Association for Cancer Research* **2009**, *15*, 452-459.

60. Yamamoto, N.; Kinoshita, T.; Nohata, N.; Yoshino, H.; Itesako, T.; Fujimura, L.; Mitsuhashi, A.; Usui, H.; Enokida, H.; Nakagawa, M., *et al.* Tumor-suppressive microrna-29a inhibits cancer cell migration and invasion via targeting hsp47 in cervical squamous cell carcinoma. *International journal of oncology* **2013**, *43*, 1855-1863.

61. Li, Y.; Wang, F.; Xu, J.; Ye, F.; Shen, Y.; Zhou, J.; Lu, W.; Wan, X.; Ma, D.; Xie, X. Progressive mirna expression profiles in cervical carcinogenesis and identification of hpv-related target genes for mir-29. *The Journal of pathology* **2011**, *224*, 484-495.

62. Gong, J.; Li, J.; Wang, Y.; Liu, C.; Jia, H.; Jiang, C.; Wang, Y.; Luo, M.; Zhao, H.; Dong, L., *et al.* Characterization of microrna-29 family expression and investigation of their mechanistic roles in gastric cancer. *Carcinogenesis* **2014**, *35*, 497-506.

63. Zhang, Q.; Stovall, D.B.; Inoue, K.; Sui, G. The oncogenic role of yin yang 1. *Critical reviews in oncogenesis* **2011**, *16*, 163-197.

64. Polager, S.; Ginsberg, D. E2f - at the crossroads of life and death. *Trends in cell biology* **2008**, *18*, 528-535.

65. Wang, X.; Meyers, C.; Guo, M.; Zheng, Z.M. Upregulation of p18ink4c expression by oncogenic hpv e6 via p53-mir-34a pathway. *International journal of cancer. Journal international du cancer* **2011**, *129*, 1362-1372.

66. Pang, R.T.; Leung, C.O.; Ye, T.M.; Liu, W.; Chiu, P.C.; Lam, K.K.; Lee, K.F.; Yeung, W.S. Microrna-34a suppresses invasion through downregulation of

notch1 and jagged1 in cervical carcinoma and choriocarcinoma cells. *Carcinogenesis* **2010**, *31*, 1037-1044.

67. Li, P.; Sheng, C.; Huang, L.; Zhang, H.; Huang, L.; Cheng, Z.; Zhu, Q. Mir-183/-96/-182 cluster is up-regulated in most breast cancers and increases cell proliferation and migration. *Breast cancer research : BCR* **2014**, *16*, 473.

68. Tang, T.; Wong, H.K.; Gu, W.; Yu, M.Y.; To, K.F.; Wang, C.C.; Wong, Y.F.; Cheung, T.H.; Chung, T.K.; Choy, K.W. Microrna-182 plays an onco-mirna role in cervical cancer. *Gynecologic oncology* **2013**, *129*, 199-208.

69. Ding, H.; Wu, Y.L.; Wang, Y.X.; Zhu, F.F. Characterization of the microrna expression profile of cervical squamous cell carcinoma metastases. *Asian Pacific journal of cancer prevention : APJCP* **2014**, *15*, 1675-1679.

70. Huang, H.; Regan, K.M.; Wang, F.; Wang, D.; Smith, D.I.; van Deursen, J.M.; Tindall, D.J. Skp2 inhibits foxo1 in tumor suppression through ubiquitin-mediated degradation. *Proceedings of the National Academy of Sciences of the United States of America* **2005**, *102*, 1649-1654.

71. Schmidt, M.; Fernandez de Mattos, S.; van der Horst, A.; Klompmaker, R.; Kops, G.J.; Lam, E.W.; Burgering, B.M.; Medema, R.H. Cell cycle inhibition by foxo forkhead transcription factors involves downregulation of cyclin d. *Molecular and cellular biology* **2002**, *22*, 7842-7852.

72. Xin, J.X.; Yue, Z.; Zhang, S.; Jiang, Z.H.; Wang, P.Y.; Li, Y.J.; Pang, M.; Xie, S.Y. Mir-99 inhibits cervical carcinoma cell proliferation by targeting trib2. *Oncology letters* **2013**, *6*, 1025-1030.

73. Grandinetti, K.B.; Stevens, T.A.; Ha, S.; Salamone, R.J.; Walker, J.R.; Zhang, J.; Agarwalla, S.; Tenen, D.G.; Peters, E.C.; Reddy, V.A. Overexpression of trib2 in human lung cancers contributes to tumorigenesis through downregulation of c/ebpalpha. *Oncogene* **2011**, *30*, 3328-3335.

74. Keeshan, K.; He, Y.; Wouters, B.J.; Shestova, O.; Xu, L.; Sai, H.; Rodriguez, C.G.; Maillard, I.; Tobias, J.W.; Valk, P., *et al.* Tribbles homolog 2 inactivates c/ebpalpha and causes acute myelogenous leukemia. *Cancer cell* **2006**, *10*, 401-411.

75. Naiki, T.; Saijou, E.; Miyaoka, Y.; Sekine, K.; Miyajima, A. Trb2, a mouse tribbles ortholog, suppresses adipocyte differentiation by inhibiting akt and c/ebpbeta. *The Journal of biological chemistry* **2007**, *282*, 24075-24082.

76. Mohankumar, K.M.; Xu, X.Q.; Zhu, T.; Kannan, N.; Miller, L.D.; Liu, E.T.; Gluckman, P.D.; Sukumar, S.; Emerald, B.S.; Lobie, P.E. Hoxa1-stimulated oncogenicity is mediated by selective upregulation of components of the p44/42 map kinase pathway in human mammary carcinoma cells. *Oncogene* **2007**, *26*, 3998-4008.

77. Chen, D.; Chen, Z.; Jin, Y.; Dragas, D.; Zhang, L.; Adjei, B.S.; Wang, A.; Dai, Y.; Zhou, X. Microrna-99 family members suppress homeobox a1 expression in epithelial cells. *PloS one* **2013**, *8*, e80625.

78. Li, B.H.; Zhou, J.S.; Ye, F.; Cheng, X.D.; Zhou, C.Y.; Lu, W.G.; Xie, X. Reduced mir-100 expression in cervical cancer and precursors and its carcinogenic effect through targeting plk1 protein. *Eur J Cancer* **2011**, *47*, 2166-2174.

79. Bahassi el, M. Polo-like kinases and DNA damage checkpoint: Beyond the traditional mitotic functions. *Exp Biol Med (Maywood)* **2011**, *236*, 648-657.

80. Hyun, S.Y.; Hwang, H.I.; Jang, Y.J. Polo-like kinase-1 in DNA damage response. *BMB reports* **2014**, *47*, 249-255.

81. Cui, F.; Li, X.; Zhu, X.; Huang, L.; Huang, Y.; Mao, C.; Yan, Q.; Zhu, J.; Zhao, W.; Shi, H. Mir-125b inhibits tumor growth and promotes apoptosis of cervical cancer cells by targeting phosphoinositide 3-kinase catalytic subunit delta. *Cellular physiology and biochemistry : international journal of experimental cellular physiology, biochemistry, and pharmacology* **2012**, *30*, 1310-1318.

82. Ke, G.; Liang, L.; Yang, J.M.; Huang, X.; Han, D.; Huang, S.; Zhao, Y.; Zha, R.; He, X.; Wu, X. Mir-181a confers resistance of cervical cancer to radiation therapy through targeting the pro-apoptotic prkcd gene. *Oncogene* **2013**, *32*, 3019-3027.

83. Chen, Y.; Ke, G.; Han, D.; Liang, S.; Yang, G.; Wu, X. Microrna-181a enhances the chemoresistance of human cervical squamous cell carcinoma to cisplatin by targeting prkcd. *Experimental cell research* **2014**, *320*, 12-20.

84. Yang, L.; Wang, Y.L.; Liu, S.; Zhang, P.P.; Chen, Z.; Liu, M.; Tang, H. Mir-181b promotes cell proliferation and reduces apoptosis by repressing the expression of adenylyl cyclase 9 (ac9) in cervical cancer cells. *FEBS letters* **2014**, *588*, 124-130.

85. Martinez-Velazquez, M.; Melendez-Zajgla, J.; Maldonado, V. Apoptosis induced by camp requires smac/diablo transcriptional upregulation. *Cellular signalling* **2007**, *19*, 1212-1220.

86. Sun, X.; Sit, A.; Feinberg, M.W. Role of mir-181 family in regulating vascular inflammation and immunity. *Trends in cardiovascular medicine* **2014**, *24*, 105-112.

87. Baumgarten, A.; Bang, C.; Tschirner, A.; Engelmann, A.; Adams, V.; von Haehling, S.; Doehner, W.; Pregla, R.; Anker, M.S.; Blecharz, K*., et al.* Twist1 regulates the activity of ubiquitin proteasome system via the mir-199/214 cluster in human end-stage dilated cardiomyopathy. *International journal of cardiology* **2013**, *168*, 1447-1452.

88. Yin, G.; Chen, R.; Alvero, A.B.; Fu, H.H.; Holmberg, J.; Glackin, C.; Rutherford, T.; Mor, G. Twisting stemness, inflammation and proliferation of epithelial ovarian cancer cells through mir199a2/214. *Oncogene* **2010**, *29*, 3545-3553.

89. Qiang, R.; Wang, F.; Shi, L.Y.; Liu, M.; Chen, S.; Wan, H.Y.; Li, Y.X.; Li, X.; Gao, S.Y.; Sun, B.C*., et al.* Plexin-b1 is a target of mir-214 in cervical cancer and promotes the growth and invasion of hela cells. *The international journal of biochemistry & cell biology* **2011**, *43*, 632-641.

90. Peng, R.Q.; Wan, H.Y.; Li, H.F.; Liu, M.; Li, X.; Tang, H. Microrna-214 suppresses growth and invasiveness of cervical cancer cells by targeting udp-n-acetyl-alpha-d-galactosamine:Polypeptide n-acetylgalactosaminyltransferase 7. *The Journal of biological chemistry* **2012**, *287*, 14301-14309.

91. Wang, F.; Liu, M.; Li, X.; Tang, H. Mir-214 reduces cell survival and enhances cisplatin-induced cytotoxicity via down-regulation of bcl2l2 in cervical cancer cells. *FEBS letters* **2013**, *587*, 488-495.

92. Cai, N.; Wang, Y.D.; Zheng, P.S. The microrna-302-367 cluster suppresses the proliferation of cervical carcinoma cells through the novel target akt1. *RNA* **2013**, *19*, 85-95.

93. Kumar, M.S.; Lu, J.; Mercer, K.L.; Golub, T.R.; Jacks, T. Impaired microrna processing enhances cellular transformation and tumorigenesis. *Nature genetics* **2007**, *39*, 673-677.

94. Lee, J.W.; Choi, C.H.; Choi, J.J.; Park, Y.A.; Kim, S.J.; Hwang, S.Y.; Kim, W.Y.; Kim, T.J.; Lee, J.H.; Kim, B.G., *et al.* Altered microrna expression in cervical carcinomas. *Clinical cancer research : an official journal of the American Association for Cancer Research* **2008**, *14*, 2535-2542.

95. Martinez, I.; Gardiner, A.S.; Board, K.F.; Monzon, F.A.; Edwards, R.P.; Khan, S.A. Human papillomavirus type 16 reduces the expression of microrna-218 in cervical carcinoma cells. *Oncogene* **2008**, *27*, 2575-2582.

96. Lui, W.O.; Pourmand, N.; Patterson, B.K.; Fire, A. Patterns of known and novel small rnas in human cervical cancer. *Cancer research* **2007**, *67*, 6031-6043.

97. Wang, X.; Tang, S.; Le, S.Y.; Lu, R.; Rader, J.S.; Meyers, C.; Zheng, Z.M. Aberrant expression of oncogenic and tumor-suppressive micrornas in cervical cancer is required for cancer cell growth. *PloS one* **2008**, *3*, e2557.

98. Reshmi, G.; Chandra, S.S.; Babu, V.J.; Babu, P.S.; Santhi, W.S.; Ramachandran, S.; Lakshmi, S.; Nair, A.S.; Pillai, M.R. Identification and analysis of novel micrornas from fragile sites of human cervical cancer: Computational and experimental approach. *Genomics* **2011**, *97*, 333-340.

99. Li, J.H.; Xiao, X.; Zhang, Y.N.; Wang, Y.M.; Feng, L.M.; Wu, Y.M.; Zhang, Y.X. Microrna mir-886-5p inhibits apoptosis by down-regulating bax expression in human cervical carcinoma cells. *Gynecologic oncology* **2011**, *120*, 145-151.

100. Wang, L.; Wang, Q.; Li, H.L.; Han, L.Y. Expression of mir200a, mir93, metastasis-related gene reck and mmp2/mmp9 in human cervical carcinoma--relationship with prognosis. *Asian Pacific journal of cancer prevention : APJCP* **2013**, *14*, 2113-2118.

101. Lajer, C.B.; Garnaes, E.; Friis-Hansen, L.; Norrild, B.; Therkildsen, M.H.; Glud, M.; Rossing, M.; Lajer, H.; Svane, D.; Skotte, L., *et al.* The role of mirnas in human papilloma virus (hpv)-associated cancers: Bridging between hpv-related head and neck cancer and cervical cancer. *British journal of cancer* **2012**, *106*, 1526-1534.

102. Muralidhar, B.; Goldstein, L.D.; Ng, G.; Winder, D.M.; Palmer, R.D.; Gooding, E.L.; Barbosa-Morais, N.L.; Mukherjee, G.; Thorne, N.P.; Roberts, I., *et al.* Global microrna profiles in cervical squamous cell carcinoma depend on drosha expression levels. *The Journal of pathology* **2007**, *212*, 368-377.

103. Muralidhar, B.; Winder, D.; Murray, M.; Palmer, R.; Barbosa-Morais, N.; Saini, H.; Roberts, I.; Pett, M.; Coleman, N. Functional evidence that drosha overexpression in cervical squamous cell carcinoma affects cell phenotype and microrna profiles. *The Journal of pathology* **2011**, *224*, 496-507.

104. Witten, D.; Tibshirani, R.; Gu, S.G.; Fire, A.; Lui, W.O. Ultra-high throughput sequencing-based small rna discovery and discrete statistical biomarker analysis in a collection of cervical tumours and matched controls. *BMC biology* **2010**, *8*, 58.

105. Liu, S.; Zhang, P.; Chen, Z.; Liu, M.; Li, X.; Tang, H. Microrna-7 downregulates xiap expression to suppress cell growth and promote apoptosis in cervical cancer cells. *FEBS letters* **2013**, *587*, 2247-2253.

106. Cheung, T.H.; Man, K.N.; Yu, M.Y.; Yim, S.F.; Siu, N.S.; Lo, K.W.; Doran, G.; Wong, R.R.; Wang, V.W.; Smith, D.I., *et al.* Dysregulated micrornas in the pathogenesis and progression of cervical neoplasm. *Cell Cycle* **2012**, *11*, 2876-2884.

107. Pereira, P.M.; Marques, J.P.; Soares, A.R.; Carreto, L.; Santos, M.A. Microrna expression variability in human cervical tissues. *PloS one* **2010**, *5*, e11780.

108. Long, M.J.; Wu, F.X.; Li, P.; Liu, M.; Li, X.; Tang, H. Microrna-10a targets chl1 and promotes cell growth, migration and invasion in human cervical cancer cells. *Cancer letters* **2012**, *324*, 186-196.

109. Ma, D.; Zhang, Y.Y.; Guo, Y.L.; Li, Z.J.; Geng, L. Profiling of microrna-mrna reveals roles of micrornas in cervical cancer. *Chinese medical journal* **2012**, *125*, 4270-4276.

110. Wang, X.; Wang, H.K.; Li, Y.; Hafner, M.; Banerjee, N.S.; Tang, S.; Briskin, D.; Meyers, C.; Chow, L.T.; Xie, X., *et al.* Micrornas are biomarkers of oncogenic human papillomavirus infections. *Proceedings of the National Academy of Sciences of the United States of America* **2014**, *111*, 4262-4267.

111. Wei, Q.; Li, Y.X.; Liu, M.; Li, X.; Tang, H. Mir-17-5p targets tp53inp1 and regulates cell proliferation and apoptosis of cervical cancer cells. *IUBMB life* **2012**, *64*, 697-704.

112. Chen, J.; Yao, D.; Li, Y.; Chen, H.; He, C.; Ding, N.; Lu, Y.; Ou, T.; Zhao, S.; Li, L., *et al.* Serum microrna expression levels can predict lymph node metastasis in patients with early-stage cervical squamous cell carcinoma. *International journal of molecular medicine* **2013**, *32*, 557-567.

113. Zhao, S.; Yao, D.S.; Chen, J.Y.; Ding, N. Aberrant expression of mir-20a and mir-203 in cervical cancer. *Asian Pacific journal of cancer prevention :* *APJCP* **2013**, *14*, 2289-2293.

114. Zhang, Y.; Dai, Y.; Huang, Y.; Ma, L.; Yin, Y.; Tang, M.; Hu, C. Microarray profile of micro-ribonucleic acid in tumor tissue from cervical squamous cell carcinoma without human papillomavirus. *The journal of obstetrics and gynaecology research* **2009**, *35*, 842-849.

115. Li, B.; Hu, Y.; Ye, F.; Li, Y.; Lv, W.; Xie, X. Reduced mir-34a expression in normal cervical tissues and cervical lesions with high-risk human papillomavirus infection. *International journal of gynecological cancer :* *official journal of the International Gynecological Cancer Society* **2010**, *20*, 597-604.

116. Liu, L.; Yu, X.; Guo, X.; Tian, Z.; Su, M.; Long, Y.; Huang, C.; Zhou, F.; Liu, M.; Wu, X., *et al.* Mir-143 is downregulated in cervical cancer and promotes apoptosis and inhibits tumor formation by targeting bcl-2. *Molecular medicine reports* **2012**, *5*, 753-760.

117. Xing, A.Y.; Wang, B.; Shi, D.B.; Zhang, X.F.; Gao, C.; He, X.Q.; Liu, W.J.; Gao, P. Deregulated expression of mir-145 in manifold human cancer cells. *Experimental and molecular pathology* **2013**, *95*, 91-97.

118. Liu, J.; Sun, H.; Wang, X.; Yu, Q.; Li, S.; Yu, X.; Gong, W. Increased exosomal microrna-21 and microrna-146a levels in the cervicovaginal lavage specimens of patients with cervical cancer. *International journal of molecular sciences* **2014**, *15*, 758-773.

119. How, C.; Hui, A.B.; Alajez, N.M.; Shi, W.; Boutros, P.C.; Clarke, B.A.; Yan, R.; Pintilie, M.; Fyles, A.; Hedley, D.W., *et al.* Microrna-196b regulates the

homeobox b7-vascular endothelial growth factor axis in cervical cancer. *PloS one* **2013**, *8*, e67846.

120. Zhu, X.; Er, K.; Mao, C.; Yan, Q.; Xu, H.; Zhang, Y.; Zhu, J.; Cui, F.; Zhao, W.; Shi, H. Mir-203 suppresses tumor growth and angiogenesis by targeting vegfa in cervical cancer. *Cellular physiology and biochemistry : international journal of experimental cellular physiology, biochemistry, and pharmacology* **2013**, *32*, 64-73.

121. Xie, H.; Zhao, Y.; Caramuta, S.; Larsson, C.; Lui, W.O. Mir-205 expression promotes cell proliferation and migration of human cervical cancer cells. *PloS one* **2012**, *7*, e46990.

122. Yu, J.; Wang, Y.; Dong, R.; Huang, X.; Ding, S.; Qiu, H. Circulating microrna-218 was reduced in cervical cancer and correlated with tumor invasion. *Journal of cancer research and clinical oncology* **2012**, *138*, 671-674.

123. Yamamoto, N.; Kinoshita, T.; Nohata, N.; Itesako, T.; Yoshino, H.; Enokida, H.; Nakagawa, M.; Shozu, M.; Seki, N. Tumor suppressive microrna-218 inhibits cancer cell migration and invasion by targeting focal adhesion pathways in cervical squamous cell carcinoma. *International journal of oncology* **2013**, *42*, 1523-1532.

124. Li, Y.; Liu, J.; Yuan, C.; Cui, B.; Zou, X.; Qiao, Y. High-risk human papillomavirus reduces the expression of microrna-218 in women with cervical intraepithelial neoplasia. *The Journal of international medical research* **2010**, *38*, 1730-1736.

125. Shen, S.N.; Wang, L.F.; Jia, Y.F.; Hao, Y.Q.; Zhang, L.; Wang, H. Upregulation of microrna-224 is associated with aggressive progression and poor prognosis in human cervical cancer. *Diagnostic pathology* **2013**, *8*, 69.

126. Bierkens, M.; Krijgsman, O.; Wilting, S.M.; Bosch, L.; Jaspers, A.; Meijer, G.A.; Meijer, C.J.; Snijders, P.J.; Ylstra, B.; Steenbergen, R.D. Focal aberrations indicate eya2 and hsa-mir-375 as oncogene and tumor suppressor in cervical carcinogenesis. *Genes, chromosomes & cancer* **2013**, *52*, 56-68.

127. Shen, Y.; Li, Y.; Ye, F.; Wang, F.; Wan, X.; Lu, W.; Xie, X. Identification of mir-23a as a novel microrna normalizer for relative quantification in human

uterine cervical tissues. *Experimental & molecular medicine* **2011**, *43*, 358-366.

128. Xu, J.; Li, Y.; Wang, F.; Wang, X.; Cheng, B.; Ye, F.; Xie, X.; Zhou, C.; Lu, W. Suppressed mir-424 expression via upregulation of target gene chk1 contributes to the progression of cervical cancer. *Oncogene* **2013**, *32*, 976-987.

129. Zhang, J.; Li, S.; Yan, Q.; Chen, X.; Yang, Y.; Liu, X.; Wan, X. Interferon-beta induced microrna-129-5p down-regulates hpv-18 e6 and e7 viral gene expression by targeting sp1 in cervical cancer cells. *PloS one* **2013**, *8*, e81366.

130. Calin, G.A.; Sevignani, C.; Dumitru, C.D.; Hyslop, T.; Noch, E.; Yendamuri, S.; Shimizu, M.; Rattan, S.; Bullrich, F.; Negrini, M., *et al.* Human microrna genes are frequently located at fragile sites and genomic regions involved in cancers. *Proceedings of the National Academy of Sciences of the United States of America* **2004**, *101*, 2999-3004.

131. Schmitz, M.; Driesch, C.; Jansen, L.; Runnebaum, I.B.; Durst, M. Non-random integration of the hpv genome in cervical cancer. *PloS one* **2012**, *7*, e39632.

132. Kraus, I.; Driesch, C.; Vinokurova, S.; Hovig, E.; Schneider, A.; von Knebel Doeberitz, M.; Durst, M. The majority of viral-cellular fusion transcripts in cervical carcinomas cotranscribe cellular sequences of known or predicted genes. *Cancer research* **2008**, *68*, 2514-2522.

133. Nuovo, G.J.; Wu, X.; Volinia, S.; Yan, F.; di Leva, G.; Chin, N.; Nicol, A.F.; Jiang, J.; Otterson, G.; Schmittgen, T.D., *et al.* Strong inverse correlation between microrna-125b and human papillomavirus DNA in productive infection. *Diagnostic molecular pathology : the American journal of surgical pathology, part B* **2010**, *19*, 135-143.

134. Soto-Reyes, E.; Gonzalez-Barrios, R.; Cisneros-Soberanis, F.; Herrera-Goepfert, R.; Perez, V.; Cantu, D.; Prada, D.; Castro, C.; Recillas-Targa, F.; Herrera, L.A. Disruption of ctcf at the mir-125b1 locus in gynecological cancers. *BMC cancer* **2012**, *12*, 40.

135. Greco, D.; Kivi, N.; Qian, K.; Leivonen, S.K.; Auvinen, P.; Auvinen, E. Human papillomavirus 16 e5 modulates the expression of host micrornas. *PloS one* **2011**, *6*, e21646.

136. Yue, C.; Wang, M.; Ding, B.; Wang, W.; Fu, S.; Zhou, D.; Zhang, Z.; Han, S. Polymorphism of the pre-mir-146a is associated with risk of cervical cancer in a chinese population. *Gynecologic oncology* **2011**, *122*, 33-37.

137. Yin, Z.; Yan, L.; Cui, Z.; Li, X.; Ren, Y.; Zhou, B. Effects of common polymorphisms rs2910164 in mir-146a and rs3746444 in mir-499 on cancer susceptibility: A meta-analysis. *Molecular biology reports* **2013**, *40*, 3003-3013.

138. Jazdzewski, K.; Murray, E.L.; Franssila, K.; Jarzab, B.; Schoenberg, D.R.; de la Chapelle, A. Common snp in pre-mir-146a decreases mature mir expression and predisposes to papillary thyroid carcinoma. *Proceedings of the National Academy of Sciences of the United States of America* **2008**, *105*, 7269-7274.

139. Brosh, R.; Shalgi, R.; Liran, A.; Landan, G.; Korotayev, K.; Nguyen, G.H.; Enerly, E.; Johnsen, H.; Buganim, Y.; Solomon, H., *et al.* P53-repressed mirnas are involved with e2f in a feed-forward loop promoting proliferation. *Molecular systems biology* **2008**, *4*, 229.

140. Zheng, Z.M.; Wang, X. Regulation of cellular mirna expression by human papillomaviruses. *Biochimica et biophysica acta* **2011**, *1809*, 668-677.

141. Cannell, I.G.; Kong, Y.W.; Johnston, S.J.; Chen, M.L.; Collins, H.M.; Dobbyn, H.C.; Elia, A.; Kress, T.R.; Dickens, M.; Clemens, M.J., *et al.* P38 mapk/mk2-mediated induction of mir-34c following DNA damage prevents myc-dependent DNA replication. *Proceedings of the National Academy of Sciences of the United States of America* **2010**, *107*, 5375-5380.

142. Bueno, M.J.; Gomez de Cedron, M.; Laresgoiti, U.; Fernandez-Piqueras, J.; Zubiaga, A.M.; Malumbres, M. Multiple e2f-induced micrornas prevent replicative stress in response to mitogenic signaling. *Molecular and cellular biology* **2010**, *30*, 2983-2995.

143. Woods, K.; Thomson, J.M.; Hammond, S.M. Direct regulation of an oncogenic micro-rna cluster by e2f transcription factors. *The Journal of biological chemistry* **2007**, *282*, 2130-2134.

144. O'Donnell, K.A.; Wentzel, E.A.; Zeller, K.I.; Dang, C.V.; Mendell, J.T. C-myc-regulated micrornas modulate e2f1 expression. *Nature* **2005**, *435*, 839-843.

145. Myklebust, M.P.; Bruland, O.; Fluge, O.; Skarstein, A.; Balteskard, L.; Dahl, O. Microrna-15b is induced with e2f-controlled genes in hpv-related cancer. *British journal of cancer* **2011**, *105*, 1719-1725.

146. Melar-New, M.; Laimins, L.A. Human papillomaviruses modulate expression of microrna 203 upon epithelial differentiation to control levels of p63 proteins. *Journal of virology* **2010**, *84*, 5212-5221.

147. He, G.; Wang, Q.; Zhou, Y.; Wu, X.; Wang, L.; Duru, N.; Kong, X.; Zhang, P.; Wan, B.; Sui, L., *et al.* Yy1 is a novel potential therapeutic target for the treatment of hpv infection-induced cervical cancer by arsenic trioxide. *International journal of gynecological cancer : official journal of the International Gynecological Cancer Society* **2011**, *21*, 1097-1104.

148. Baldwin, A.; Li, W.; Grace, M.; Pearlberg, J.; Harlow, E.; Munger, K.; Grueneberg, D.A. Kinase requirements in human cells: Ii. Genetic interaction screens identify kinase requirements following hpv16 e7 expression in cancer cells. *Proceedings of the National Academy of Sciences of the United States of America* **2008**, *105*, 16478-16483.

149. Bolos, V.; Grego-Bessa, J.; de la Pompa, J.L. Notch signaling in development and cancer. *Endocrine reviews* **2007**, *28*, 339-363.

150. Zhang, Y.; Liu, Y.; Yang, Y.X.; Xia, J.H.; Zhang, H.X.; Li, H.B.; Yu, C.Z. The expression of plk-1 in cervical carcinoma: A possible target for enhancing chemosensitivity. *Journal of experimental & clinical cancer research : CR* **2009**, *28*, 130.

151. Pillai, M.R.; Halabi, S.; McKalip, A.; Jayaprakash, P.G.; Rajalekshmi, T.N.; Nair, M.K.; Herman, B. The presence of human papillomavirus-16/-18 e6, p53, and bcl-2 protein in cervicovaginal smears from patients with invasive cervical cancer. *Cancer epidemiology, biomarkers & prevention : a publication of the American Association for Cancer Research, cosponsored by the American Society of Preventive Oncology* **1996**, *5*, 329-335.

152. Dimitrakakis, C.; Kymionis, G.; Diakomanolis, E.; Papaspyrou, I.; Rodolakis, A.; Arzimanoglou, I.; Leandros, E.; Michalas, S. The possible role of p53 and bcl-2 expression in cervical carcinomas and their premalignant lesions. *Gynecologic oncology* **2000**, *77*, 129-136.

153. Baserga, R. The insulin receptor substrate-1: A biomarker for cancer? *Experimental cell research* **2009**, *315*, 727-732.

154. Wei, X.; Xu, H.; Kufe, D. Human mucin 1 oncoprotein represses transcription of the p53 tumor suppressor gene. *Cancer research* **2007**, *67*, 1853-1858.

155. Shi, M.; Du, L.; Liu, D.; Qian, L.; Hu, M.; Yu, M.; Yang, Z.; Zhao, M.; Chen, C.; Guo, L., *et al.* Glucocorticoid regulation of a novel hpv-e6-p53-mir-145 pathway modulates invasion and therapy resistance of cervical cancer cells. *The Journal of pathology* **2012**, *228*, 148-157.

156. Shi, T.Y.; Chen, X.J.; Zhu, M.L.; Wang, M.Y.; He, J.; Yu, K.D.; Shao, Z.M.; Sun, M.H.; Zhou, X.Y.; Cheng, X., *et al.* A pri-mir-218 variant and risk of cervical carcinoma in chinese women. *BMC cancer* **2013**, *13*, 19.

157. Alajez, N.M.; Lenarduzzi, M.; Ito, E.; Hui, A.B.; Shi, W.; Bruce, J.; Yue, S.; Huang, S.H.; Xu, W.; Waldron, J., *et al.* Mir-218 suppresses nasopharyngeal cancer progression through downregulation of survivin and the slit2-robo1 pathway. *Cancer research* **2011**, *71*, 2381-2391.

158. Wang, Q.; Shu, R.; He, H.; Wang, L.; Ma, Y.; Zhu, H.; Wang, Z.; Wang, S.; Shen, G.; Lei, P. Co-silencing of birc5 (survivin) and hspa5 (grp78) induces apoptosis in hepatoma cells more efficiently than single gene interference. *International journal of oncology* **2012**, *41*, 652-660.

159. Sukpan, K.; Settakorn, J.; Khunamornpong, S.; Cheewakriangkrai, C.; Srisomboon, J.; Siriaunkgul, S. Expression of survivin, cd117, and c-erbb-2 in neuroendocrine carcinoma of the uterine cervix. *International journal of gynecological cancer : official journal of the International Gynecological Cancer Society* **2011**, *21*, 911-917.

160. Narayan, G.; Goparaju, C.; Arias-Pulido, H.; Kaufmann, A.M.; Schneider, A.; Durst, M.; Mansukhani, M.; Pothuri, B.; Murty, V.V. Promoter hypermethylation-mediated inactivation of multiple slit-robo pathway genes in cervical cancer progression. *Molecular cancer* **2006**, *5*, 16.

161. Yi, S.; Chen, Y.; Wen, L.; Yang, L.; Cui, G. Expression of connexin 32 and connexin 43 in acute myeloid leukemia and their roles in proliferation. *Oncology letters* **2012**, *4*, 1003-1007.

162. Macdonald, A.I.; Sun, P.; Hernandez-Lopez, H.; Aasen, T.; Hodgins, M.B.; Edward, M.; Roberts, S.; Massimi, P.; Thomas, M.; Banks, L., *et al.* A functional interaction between the maguk protein hdlg and the gap junction protein connexin 43 in cervical tumour cells. *The Biochemical journal* **2012**, *446*, 9-21.

163. Skyldberg, B.; Salo, S.; Eriksson, E.; Aspenblad, U.; Moberger, B.; Tryggvason, K.; Auer, G. Laminin-5 as a marker of invasiveness in cervical lesions. *Journal of the National Cancer Institute* **1999**, *91*, 1882-1887.

164. Kohlberger, P.; Beneder, C.; Horvat, R.; Leodolter, S.; Breitenecker, G. Immunohistochemical expression of laminin-5 in cervical intraepithelial neoplasia. *Gynecologic oncology* **2003**, *89*, 391-394.

165. Li, J.; Ping, Z.; Ning, H. Mir-218 impairs tumor growth and increases chemosensitivity to cisplatin in cervical cancer. *International journal of molecular sciences* **2012**, *13*, 16053-16064.

166. Brehm, A.; Nielsen, S.J.; Miska, E.A.; McCance, D.J.; Reid, J.L.; Bannister, A.J.; Kouzarides, T. The e7 oncoprotein associates with mi2 and histone deacetylase activity to promote cell growth. *The EMBO journal* **1999**, *18*, 2449-2458.

167. Longworth, M.S.; Wilson, R.; Laimins, L.A. Hpv31 e7 facilitates replication by activating e2f2 transcription through its interaction with hdacs. *The EMBO journal* **2005**, *24*, 1821-1830.

168. Wada, T.; Kikuchi, J.; Furukawa, Y. Histone deacetylase 1 enhances microrna processing via deacetylation of dgcr8. *EMBO reports* **2012**, *13*, 142-149.

169. Wang, F.; Li, Y.; Zhou, J.; Xu, J.; Peng, C.; Ye, F.; Shen, Y.; Lu, W.; Wan, X.; Xie, X. Mir-375 is down-regulated in squamous cervical cancer and inhibits cell migration and invasion via targeting transcription factor sp1. *The American journal of pathology* **2011**, *179*, 2580-2588.

170. Peralta-Zaragoza, O.; Bermudez-Morales, V.; Gutierrez-Xicotencatl, L.; Alcocer-Gonzalez, J.; Recillas-Targa, F.; Madrid-Marina, V. E6 and e7 oncoproteins from human papillomavirus type 16 induce activation of human transforming growth factor beta1 promoter throughout sp1 recognition sequence. *Viral immunology* **2006**, *19*, 468-480.

171. Lee, Y.J.; Lee, J.E.; Choi, H.J.; Lim, J.S.; Jung, H.J.; Baek, M.C.; Frokiaer, J.; Nielsen, S.; Kwon, T.H. E3 ubiquitin-protein ligases in rat kidney collecting duct: Response to vasopressin stimulation and withdrawal. *American journal of physiology. Renal physiology* **2011**, *301*, F883-896.

172. Feng, L.; Allen, N.S.; Simo, S.; Cooper, J.A. Cullin 5 regulates dab1 protein levels and neuron positioning during cortical development. *Genes & development* **2007**, *21*, 2717-2730.

173. Lewis, S.P.; Willis, A.N.; Johnson, A.E.; Resau, J.; Burnatowska-Hledin, M.A. Mutational analysis of vacm-1/cul5 exons in cancer cell lines. *APMIS : acta pathologica, microbiologica, et immunologica Scandinavica* **2011**, *119*, 421-430.

174. Lankat-Buttgereit, B.; Goke, R. The tumour suppressor pdcd4: Recent advances in the elucidation of function and regulation. *Biology of the cell / under the auspices of the European Cell Biology Organization* **2009**, *101*, 309-317.

175. Yao, Q.; Xu, H.; Zhang, Q.Q.; Zhou, H.; Qu, L.H. Microrna-21 promotes cell proliferation and down-regulates the expression of programmed cell death 4 (pdcd4) in hela cervical carcinoma cells. *Biochemical and biophysical research communications* **2009**, *388*, 539-542.

176. Deftereos, G.; Corrie, S.R.; Feng, Q.; Morihara, J.; Stern, J.; Hawes, S.E.; Kiviat, N.B. Expression of mir-21 and mir-143 in cervical specimens ranging from histologically normal through to invasive cervical cancer. *PloS one* **2011**, *6*, e28423.

177. Lei, C.; Wang, Y.; Huang, Y.; Yu, H.; Huang, Y.; Wu, L.; Huang, L. Up-regulated mir155 reverses the epithelial-mesenchymal transition induced by egf and increases chemo-sensitivity to cisplatin in human caski cervical cancer cells. *PloS one* **2012**, *7*, e52310.

178. Wan, G.; Xie, W.; Liu, Z.; Xu, W.; Lao, Y.; Huang, N.; Cui, K.; Liao, M.; He, J.; Jiang, Y., *et al.* Hypoxia-induced mir155 is a potent autophagy inducer by targeting multiple players in the mtor pathway. *Autophagy* **2014**, *10*, 70-79.

179. Wilting, S.M.; van Boerdonk, R.A.; Henken, F.E.; Meijer, C.J.; Diosdado, B.; Meijer, G.A.; le Sage, C.; Agami, R.; Snijders, P.J.; Steenbergen, R.D.

Methylation-mediated silencing and tumour suppressive function of hsa-mir-124 in cervical cancer. *Molecular cancer* **2010**, *9*, 167.

180. Harris, T.G.; Burk, R.D.; Yu, H.; Minkoff, H.; Massad, L.S.; Watts, D.H.; Zhong, Y.; Gange, S.; Kaplan, R.C.; Anastos, K., *et al.* Insulin-like growth factor axis and oncogenic human papillomavirus natural history. *Cancer epidemiology, biomarkers & prevention : a publication of the American Association for Cancer Research, cosponsored by the American Society of Preventive Oncology* **2008**, *17*, 245-248.

181. Hirano, S.; Ito, N.; Takahashi, S.; Tamaya, T. Clinical implications of insulin-like growth factors through the presence of their binding proteins and receptors expressed in gynecological cancers. *European journal of gynaecological oncology* **2004**, *25*, 187-191.

182. Tomasini, R.; Samir, A.A.; Carrier, A.; Isnardon, D.; Cecchinelli, B.; Soddu, S.; Malissen, B.; Dagorn, J.C.; Iovanna, J.L.; Dusetti, N.J. Tp53inp1s and homeodomain-interacting protein kinase-2 (hipk2) are partners in regulating p53 activity. *The Journal of biological chemistry* **2003**, *278*, 37722-37729.

183. Yan, H.L.; Xue, G.; Mei, Q.; Wang, Y.Z.; Ding, F.X.; Liu, M.F.; Lu, M.H.; Tang, Y.; Yu, H.Y.; Sun, S.H. Repression of the mir-17-92 cluster by p53 has an important function in hypoxia-induced apoptosis. *The EMBO journal* **2009**, *28*, 2719-2732.

184. Wang, Y.D.; Cai, N.; Wu, X.L.; Cao, H.Z.; Xie, L.L.; Zheng, P.S. Oct4 promotes tumorigenesis and inhibits apoptosis of cervical cancer cells by mir-125b/bak1 pathway. *Cell death & disease* **2013**, *4*, e760.

185. Brockhausen, I. Pathways of o-glycan biosynthesis in cancer cells. *Biochimica et biophysica acta* **1999**, *1473*, 67-95.

186. Vasudevan, S.; Tong, Y.; Steitz, J.A. Switching from repression to activation: Micrornas can up-regulate translation. *Science* **2007**, *318*, 1931-1934.

187. Noland, C.L.; Doudna, J.A. Multiple sensors ensure guide strand selection in human rnai pathways. *RNA* **2013**, *19*, 639-648.

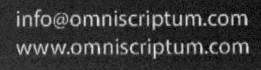

Printed by Books on Demand GmbH, Norderstedt / Germany